Mangrove Biology, Ecosystem, and Conservation

Edited by Orlex Baylen Yllano

Published in London, United Kingdom

Mangrove Biology, Ecosystem, and Conservation
http://dx.doi.org/10.5772/intechopen.104370
Edited by Orlex Baylen Yllano

Contributors
I. Ketut Ginantra, Izuchukwu Uche, David Alfaro Siqueiros Beltrones, Anthony E. Ogbeibu, Blessing J. Oribhabor, Klaus Schmitt, Thorsten Albers, Hamet Diaw Diadhiou, Moustapha Deme, Djiby Thiam, Maulinna Kusumo Wardhani

First published in London, United Kingdom, 2023 by IntechOpen
IntechOpen is the global imprint of INTECHOPEN LIMITED, registered in England and Wales, registration number: 11086078, 5 Princes Gate Court, London, SW7 2QJ, United Kingdom

British Library Cataloguing-in-Publication Data
A catalogue record for this book is available from the British Library

Additional hard and PDF copies can be obtained from orders@intechopen.com

Mangrove Biology, Ecosystem, and Conservation
Edited by Orlex Baylen Yllano
p. cm.

This title is part of the Environmental Sciences Book Series, Volume 11
Topic: Ecosystems and Biodiversity
Series Editor: J. Kevin Summers
Topic Editors: Salustiano Mato, Josefina Garrido and Francisco Ramil

Print ISBN 978-1-83768-589-9
Online ISBN 978-1-83768-590-5
eBook (PDF) ISBN 978-1-83768-591-2
ISSN 2754-6713

For EU product safety concerns:
IN TECH d.o.o., Prolaz Marije Krucifikse Kozulić 3, 51000 Rijeka, Croatia,
info@intechopen.com or visit our website at intechopen.com.

IntechOpen Book Series

Environmental Sciences

Volume 11

Aims and Scope of the Series

Scientists have long researched to understand the environment and man's place in it. The search for this knowledge grows in importance as rapid increases in population and economic development intensify humans' stresses on ecosystems. Fortunately, rapid increases in multiple scientific areas are advancing our understanding of environmental sciences. Breakthroughs in computing, molecular biology, ecology, and sustainability science are enhancing our ability to utilize environmental sciences to address real-world problems.

The four topics of this book series - Pollution; Environmental Resilience and Management; Ecosystems and Biodiversity; and Water Science - will address important areas of advancement in the environmental sciences. They will represent an excellent initial grouping of published works on these critical topics.

Meet the Series Editor

J. Kevin Summers is a Senior Research Ecologist at the Environmental Protection Agency's (EPA) Gulf Ecosystem Measurement and Modeling Division. He is currently working with colleagues in the Sustainable and Healthy Communities Program to develop an index of community resilience to natural hazards, an index of human well-being that can be linked to changes in the ecosystem, social and economic services, and a community sustainability tool for communities with populations under 40,000. He leads research efforts for indicator and indices development. Dr. Summers is a systems ecologist and began his career at the EPA in 1989 and has worked in various programs and capacities. This includes leading the National Coastal Assessment in collaboration with the Office of Water which culminated in the award-winning National Coastal Condition Report series (four volumes between 2001 and 2012), and which integrates water quality, sediment quality, habitat, and biological data to assess the ecosystem condition of the United States estuaries. He was acting National Program Director for Ecology for the EPA between 2004 and 2006. He has authored approximately 150 peer-reviewed journal articles, book chapters, and reports and has received many awards for technical accomplishments from the EPA and from outside of the agency. Dr. Summers holds a BA in Zoology and Psychology, an MA in Ecology, and Ph.D. in Systems Ecology/Biology.

Meet the Volume Editor

Dr. Orlex Baylen Yllano is a professor in the Department of Biology, College of Science and Technology, Adventist University of the Philippines. He belongs to a dynamic group of international researchers who have conducted mangrove research in the Indo-West Pacific region. Dr. Yllano has published papers on mangrove ecology, phytoremediation, and conservation genetics. He has mentored graduate and undergraduate thesis students. He has published research papers in Scopus/ISI-indexed journals and co-authored several books. He is affiliated with national and international scientific societies. He has served as an editorial board member and reviewer for journals. He is a recipient of several research and academic awards.

Contents

Preface

Mangroves' remarkable biology and ecosystem are vital to our planet's health and well-being, demanding our unwavering commitment to conservation. The mangrove ecosystem provides indispensable ecological services, strengthening coastal resilience, promoting biodiversity, and underlining the urgent need for protection.

Chapter 1, "Mangroves of the Niger Delta", by Izuchukwu Uche, and Chapter 2, "The Niger Delta Mangrove Ecosystem and Its Conservation Challenges", by Anthony E. Ogbeibu and Blessing J. Oribhabor, explore the rich and extensive mangroves in Africa, which provide ecosystem services to the community, outline the threats to the mangrove ecosystem, and recommend strategies and conservation efforts to sustain the Niger Delta mangrove forest.

Mangroves' growth, development, and proliferation greatly depend on their biophysicochemical substrate. Chapter 3, "Perspective Chapter: Remarks on the Relationship of Mangrove Recruitment and Thrombolithic Development in Coastal Lagoons", by David Alfaro Siqueiros Beltrones, documents the thrombolithic formations associated with mangrove forest substrate. The findings presented in this study shed new light on the intricate relationship between mangrove recruitment and the development of thrombolytic formations, which play a vital role in the geomorphological dynamics of coastal areas.

Despite efforts to conserve the fragile mangrove ecosystem, this unique environment is constantly influenced by anthropogenic activities. Chapter 4, "Perspective Chapter: Mangrove Deforestation and Sustainability of Malacological Resource Exploitation in the Estuarine Ecosystem of the Saloum River Delta, Senegal", by Hamet Diaw Diadhiou, Moustapha Deme, and Djiby Thiam, provides insights into the anthropogenic activities and sustainability of malacological resources in the estuarine ecosystem.

Proper management of mangroves greatly depends on the periodic health assessment of the ecosystem. Chapter 5, "Mangrove Health Assessment Using Hemispherical Photography: A Case Study on Mangrove Ecosystem for Ecotourism at Tajungan-Bangkalan, Madura Island, Indonesia", by Maulinna Kusumo Wardhani, highlights the critical role of mangrove health assessment for the sustainable use and management of the ecosystem. Using hemispherical photographic analyses, the observation of mangrove percent cover, monitoring, and assessment become more efficient. The findings of the study are invaluable in mangrove rehabilitation, conservation, and educational ecotourism.

Ecosystem-based coastal protection is integral to the success of the mangrove development program. Chapter 6, "Mangroves and Ecosystem-Based Coastal Protection in the Mekong River Delta, Vietnam", by Klaus Schmitt and Thorsten Albers, explores the anthropogenic and natural threats to the mangrove ecosystem in the Mekong River Delta. The chapter also highlights the ingenious and cost-effective strategies to reduce erosion and restore tidal flats for mangrove regeneration.

Finally, Chapter 7, "Perspective Chapter: Mangrove Conservation – An Ecotourism Approach", by I. Ketut Ginantra, investigates the uniqueness and diversity of the mangrove ecosystem. This ecosystem's diverse flora and fauna are platforms for scientific exploration and ecotourism. Principles and perspectives on mangrove conservation, ecology, economy, and community evaluation are well presented.

I sincerely thank all authors for sharing their esteemed works. I hope this book will inspire readers to explore and protect the astounding mangroves of the world.

Dr. Orlex Baylen Yllano
Professor and Chair,
Department of Biology,
College of Science and Technology,
Adventist University of the Philippines,
Silang, Cavite, Philippines

Chapter 1

Mangroves of the Niger Delta

Izuchukwu Uche

Abstract

The Niger Delta is located in the southern part of Nigeria; three core states, Rivers, Bayelsa, and Delta house the largest concentration of Mangroves. The Niger Delta has the most extensive mangroves in Africa; six true mangrove species are found in the Niger Delta, *Rhizophora spp*. (*R. mangle, R. harrisonii, R. racemosa*) Known as red mangrove; Avicenna *germinans,* known as black mangrove*; Langucularia racemosa,* known as white mangrove*; and Conocarpus erectus,* known as buttonwood. *Rhizophora spp.* is the dominant mangrove in the Niger Delta and constitutes over 90% of the entire mangrove species in the region. Niger Delta mangroves are essential in providing ecosystem goods and services to the people. Unfortunately, the Niger Mangrove ecosystem faces severe threats from crude oil spills resulting from equipment failure, bunkering, vandalisation, and illegal refining. The invasion of mangroves by Nipa palm in the Niger Delta has become a threat to the mangrove ecosystem; deforestation and urbanisation are also significant threats affecting mangroves in the region. Providing sustainable alternatives to mangrove forest wood, conservation of mangrove forests, and enforcement of strict regulations in the oil and gas industry are some measures to ensure the sustainability of the Niger Delta mangrove forest.

Keywords: Niger Delta, mangrove, oil spill, Nipa palm, ecosystem, remediation, forest

1. Introduction

The Niger Delta region of Nigeria is home to Africa's largest mangrove forest and one of the world's most extensive mangrove forests. The area of mangrove forest habitat coverage in Nigeria is estimated at 8442.43 km^2 [1]. The Niger Delta region is located in the southern part of Nigeria; this region contains the majority of the mangrove forest in the country. Mangroves are trees and shrubs that grow on intertidal coastlines of tropical and subtropical regions. Mangroves are salt-tolerant plants that survive mostly in brackish water. In the Niger Delta, four mangrove genera exist; Rhizophora, Aviecina, Langucularia, and Conocarpus. *Rhizophora spp.* include *R. racemosa, R. mangle and R. harrisonni*. Other mangrove species in the Niger Delta are *Langucularia racemosa*, *Avicennia germinans*, and *Conocarpus erectus*. *Rhizophora sp. Langucularia racemosa and Aviecina germinans* are also known as red, white, and black mangroves, respectively, while *C. erectus* is known as buttonwood in the Niger Delta [2, 3]. Red mangroves are easily identified in the Niger Delta by their prop roots, leaves, and propagules; white mangroves are easily identified by their oval leaves, fruits, and seeds; and black mangroves can easily be distinguished from other mangroves by their elongated leaves, fruits, and seeds, including the presence of pneumatophores.

Mangroves in the region have been severely depleted; just like mangroves worldwide, mangroves in the Niger Delta are being lost at an alarming rate. However, data are readily not available for the hectares of mangrove forests lost over the last two decades in the Niger Delta. Several researchers have reported significant losses. Today's leading cause of mangrove loss in the Niger Delta is crude oil spills and artisanal refining [4]. Other factors have also been reported to cause the loss of mangrove forests in the region: urbanisation, conversion of mangrove forests for agriculture and aquaculture, construction, deforestation, overpopulation, and the invasion of nipa palm in the mangrove ecosystem. The importance of the mangrove ecosystem has been widely researched and documented, so the need to protect and preserve these ecosystems becomes very dire owing to their uniqueness and myriads of ecosystem goods and services [5].

While the degradation and destruction of mangrove forests in Africa's most extensive mangrove continues, efforts have been made to restore hectares of mangrove forest destroyed by a crude oil spill. One such effort is a litigation case between a small community in River's state and an international oil and gas company operating in the region, where an agreement was reached between the community and the international oil and gas company to clean up and revegetate approximately 1000 ha of degraded mangrove forest in the community. Furthermore, the United Nations Environmental Program (UNEP) report [6] led a government agency to clean up the oil spill in Ogoniland and revegetate damaged mangrove forests, among other functions. While clean-up and revegetation efforts have commenced in the earlier case study, clean-up is currently ongoing in the government-managed cleanup process in Ogoniland [7, 8].

2. The Niger Delta

The Niger Delta of Nigeria is among the largest delta in the world. A delta is a landform that originates due to depositions of sediments carried by a river as the flow leaves the mouth of the river and enters slower-standing or moving water. Deltas occur when a river joins a sea or an ocean and cannot transport away the supplied sediments. The Niger Delta region has the largest wetland in Africa and the third-largest wetland in the world. Projections of the estimate of the current population in the region at a growth rate of 2.9% place the population of the Niger Delta as of 2022 to be well over 45 million people. The core Delta areas in the region lie in three central states: Delta, Rivers, and Bayelsa; these three states hold the most significant amount of mangrove forest in the country. Politically the Niger Delta region in Nigeria comprises nine states: Abia, Akwa Ibom, Cross River, Edo, Imo, and Ondo states, including the core states of the Niger Delta [9]. **Figure 1** shows the map of Nigeria and the extent of coverage and distribution of mangrove forests in the Niger Delta and Nigeria.

Other states with mangroves in Nigeria include Akwa Ibom, Cross River, Ondo, Ogun, and Lagos. The mangroves of the Niger Delta are located in the lower tidal floodplain, which also comprises various creeks and estuaries. The delta has a very high rainfall of more than 2500 mm per year, with a temperature range from 18 to 30°C. The tide in the mangrove forest is diurnal and can reach an amplitude of 2.8 m during spring tide and 0.2 m at the shallow tide. The poverty rate in the Niger Delta is at an average of 30%, and the region is densely populated with a literacy rate of 70%. Most rural dwellers are into fishing and farming and mostly depend on forests and surrounding water bodies for their primary source of income and survival [9].

DOI: http://dx.doi.org/10.5772/intechopen.109296

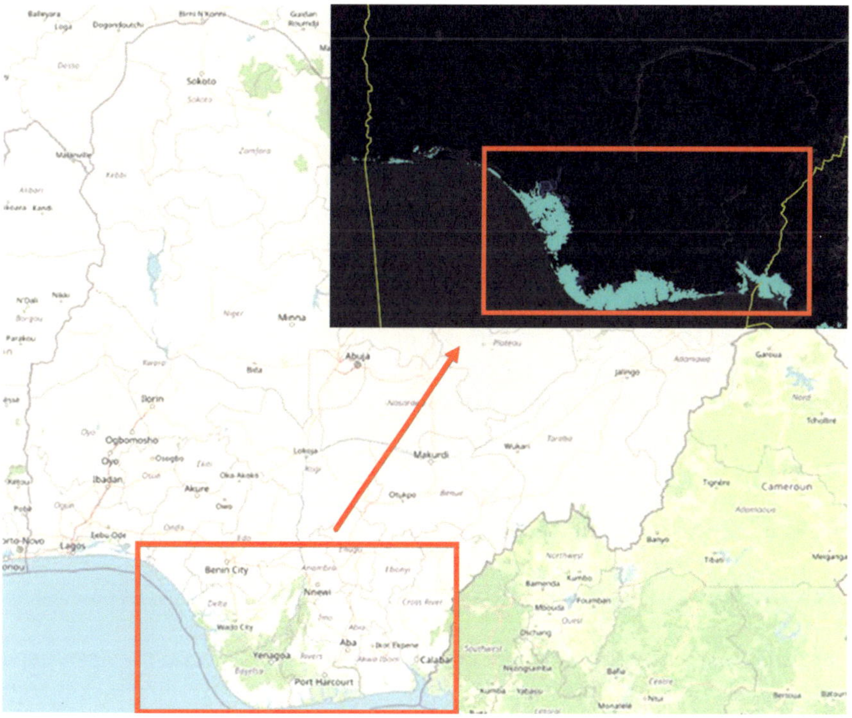

Figure 1.
Map of Nigeria showing areas of mangrove coverage in the Niger Delta and Nigeria.

The Niger Delta is also an extremely prolific hydrocarbon province where oil and gas have been explored and exploited over decades. A significant percentage of Nigeria's oil and gas facilities run through the mangrove forest in the region; this has led to the rapidly under-reported degradation of mangroves in the region either by direct or indirect consequences of oil and gas exploration. Crude oil spills resulting from equipment failure, sabotage, bunkering, and artisanal (illegal) refining in the region have destroyed vast hectares of mangrove forest in the entire Niger Delta. Although the extent of the destruction is not known throughout the entire region, significant damage to the mangrove forest has been reported in the past by researchers [4, 10, 11].

3. Mangrove autecology in the Niger Delta

3.1 Red mangrove

Rhizophora spp. is the dominant mangrove species in the Niger Delta; it is distributed on all intertidal mangrove platforms and is found in almost all forest zones in the region. Mangrove platforms are areas where mangroves thrive; platforms are mostly intertidal areas in a mangrove forest. *Rhizophora spp*. Consist of over 90% of the total mangrove plant in the Niger Delta forest. Three species of Rhizophora exist in the Niger Delta: *Rhizophora racemosa, Rhizophora mangle, and Rhizophora harrisonii*. The differences between *Rhizophora* species in the Niger Delta can mainly be observed when the plants are fruiting. These distinctions are only observable once the plants mature enough to bear fruits. For this reason, it is not easy to differentiate the various

Rhizophora species. *Rhizophora racemosa* is dominant among the three species, according to [2, 3]; in some instances, a different *Rhizophora* species is dominant in a different mangrove forest in the Niger Delta. Rhizophora mangroves have been known to grow as tall as 25-30 m in a pristine mangrove forest. They can be clearly distinguished by their prop root, which drops down from tree trunks and branches as high as 10 m in some plants, and the prop root mostly touches the soil or can be suspended in the air. Their prop roots have tiny pores all over them, which are used for gaseous exchange; these pores are known as lenticels. All year round, red mangroves in the Niger Delta flower exhibit a form of reproduction known as vivipary, where flowers develop into seedlings while still attached to the parent plant. Rhizophora seedlings are often referred to as propagules. Rhizophora propagules are distinct from other species in the Niger Delta. Thus, due to their dominance, most people confuse them for being the only mangrove species in the Niger Delta. Rhizophora propagules fall off the parent plant upon maturity and are carried by tidal currents until it finds suitable soil where it roots itself and begins to grow. Red mangroves are the most cultivated mangrove in nurseries all over the region (**Figure 2**). Shows various life stages of a typical red mangrove plant in the Niger Delta, its prop root, and propagules.

3.2 Black and white mangrove

White and black mangroves make up less than 10% of the total mangrove species population in the region. White mangroves clearly distinguish from black mangroves in the Niger Delta as seen from their leaves. The leaves of white mangroves are oval in shape with a thick, dark green colouration in most cases, while the leaves

Figure 2.
Shows a typical red mangrove plant in the Niger Delta, and its various life stages.

of black mangroves are elongated and attenuated with a light green colouration. Also, one common feature of black mangroves that distinguishes them from white mangroves is the presence of pneumatophores. Pneumatophores are the lateral root that grows out of the soil, also used for gaseous exchange by black mangrove plants. Pneumatophores are found on all black mangrove plants, extending as far as 10 m from the parent plants in matured black mangrove plants. White mangroves can grow into trees but are mainly shrubs in the Niger Delta; some white mangrove plants have pneumathodes similar to black mangrove pneumatophores but are not the same and are not found on all white mangrove plants in the region. In the Niger Delta, white and black mangroves are always found in the intertidal zones of the mangrove forest.

In contrast, black mangroves are majorly found along shorelines where the sediments contain lots of sand; black mangroves are also found on intertidal platforms where the sediments are dominated by mud. White mangroves also grow on muddy and sandy sediments like black mangroves, but they also thrive on the fringes of platforms close to the channel edges [3]. **Figure 3** shows black and white mangrove plants and their different life stages.

3.3 Buttonwood

Buttonwood is the least common mangrove species in the Niger Delta, and they are less than 1% of the total mangrove population in the region. Buttonwood grows majorly high up the platform in the mangrove forest of the Niger Delta; they are found in areas close to the high tide swash line on the platform; buttonwood does not grow on intertidal zones in the Niger Delta and grows best on sandy sediments. Their leaves are elongated and attenuated with dark green colouration, and their seeds are round and green but turn brownish red upon maturity. **Figure 4** shows a buttonwood mangrove plant in the Niger Delta and its seeds (propagule).

Figure 3.
Is black and white mangrove plants showing their various life stages.

Figure 4.
Shows a buttonwood mangrove plant in the Niger Delta and its seeds.

3.4 Mangrove associate species in the Niger Delta

Nipa palm, although an exotic species in the region, is also referred to as a mangrove associate plant despite the plant invasion of the mangrove forest in the region. Other commonly found associate mangrove species found in the region's mangrove forest are mangrove fern (*Achrostichum aureum*), mangrove grass (*Paspalum vaginatum*), and wild palm (*Phoenix reclinata*). All later species mentioned are not invasive but can be found in a different zone and intertidal areas in a mangrove forest. Mangrove fern is evenly distributed on almost all mangrove platforms; mangrove grass is mainly found on shorelines, sandy sediments, and areas close to the high tide swash line along the mangrove platform. The wild palm grows along the buttonwood corridor but is sometimes found on the intertidal platform. Wild palm and mangrove ferns grow mainly as shrubs in the region. **Figure 5** shows mangrove fern and wild palm in a mangrove forest in River's state.

3.5 Mangrove zonation in the Niger Delta

Figure 6 below is a drone picture of a remediated mangrove forest platform with little vegetation. The arrows indicate areas where each mangrove species will likely grow in the Niger Delta. The yellow arrow indicates the intertidal zone of the platform, and the yellow arrowhead to the right is the high tide swash line. In a pristine mangrove forest, due to the dominance of red mangroves, they mostly occupy the intertidal zones, as noted by the red arrow in the picture below. White mangrove also thrives in areas where red mangrove thrives but are outcompeted by red mangrove, as seen from the white arrow in the picture. Black mangroves thrive best at the upper platforms and shorelines, as indicated by the black arrow in the picture. Buttonwood typically grows in areas outside the intertidal zone or close to the mangrove high tide swash line, or it can grow in higher platforms with minor inundation. In a degraded mangrove forest, black, white, and red mangroves can thrive in any intertidal area in a mangrove platform in the Niger Delta.

3.6 Sediment types in a typical Niger Delta mangrove forest

The sediment in a typical Niger Delta mangrove forest varies from platform fringes, which are usually composed of pure mud, down to the intertidal

Figure 5.
Shows mangrove fern and wild palm common mangrove associate plant in the Niger Delta.

Figure 6.
Below is a drone picture of a remediated mangrove forest platform with little vegetation. The arrows indicate areas where each mangrove species will likely grow in the Niger Delta.

platform, where a mix of mud and Chikoko roots start to build up, down to the inner fringe of the platform and shoreline, which consist of mud, Chikoko root and sand. Chikoko roots are the thick fibrous peaty composition of mangrove roots, leaves, and other organic components of mangrove plants that have decayed over a long period and are often found in the sediments of mangrove soils. In some mangrove platforms, these fibrous peaty materials are found as deep as 1 m when you dig the soil of the mangrove platform. In certain areas, Chikoko roots make up over 95% of the entire soil sediment in the Niger Delta. Sandy soil is a significant sediment found in mangrove platforms and often along the shoreline and areas close to the high tide swash line on the platform. Soil sediments also vary from one location to the other in the Niger Delta; specific platforms in some mangrove areas are composed of two or more sediment types in different proportions. Typically, muddy sediment is the dominant sediment type in Niger Delta; it is common to see a mixture of two or more sediment types mixed or in layers. A mixture of mud and Chikoko root is a common sight when you dig up to 30 cm beneath the surface [12]. **Figure 7** below shows various sediment types commonly found in the Niger Delta mangrove forest 0.3 m below the surface sediment.

Figure 7.
Below shows various sediment types commonly found in the Niger Delta mangrove forest at a depth of 30cm.

4. Threats to Niger Delta mangrove forest

Over the years, mangroves in the Niger Delta have faced challenges common to mangroves forest worldwide: the conversion of mangrove habitats for agriculture and aquaculture. It is a common site in most mangrove areas in the Niger Delta for locals to construct fish ponds, where they rear fish and use such areas sometimes to trap fish at high tide. Areas converted to fish ponds are mainly mangrove forests cleared and used as fish ponds; in some instances, these fish ponds can be as large as 900sqm. Fishing in the Niger Delta is one of the biggest employers of labour for both men, women, and children living near mangrove habitats. Mangrove forests are cleared so nets can be used to set traps for fish; However, fish pond construction completely clears mangrove trees on the land where it is constructed; the surrounding mangrove environment is usually left undisturbed; the impact of the fish pond on the degradation of the Niger

Delta mangroves forest can be said to be minor when you assess the risk of fish pond construction in a mangrove forest in the Niger Delta.

Deforestation of mangrove forests is also a significant threat to mangroves in the Niger Delta. In the Niger Delta today, due to the high poverty and illiteracy level, most communities still use wood as their primary energy source. This has led to the constant deforestation of mangrove forests in the region. Mangrove trees make good wood for local consumption; logging trees in mangrove forests for commercial and subsistence purposes is a common site in the region. The need for cheaper energy sources and lack of alternative cheaper energy for cooking in the Niger Delta have been the reason for the constant deforestation of the Niger Delta mangrove forest. It is also common to see fishermen use deforested trees as sails for their boats. Deforestation has increased due to harsh economic realities and rising poverty levels, which have increased in the Niger Delta in the last decade. Steps must be taken to enlighten the locals on the dangers of deforesting mangrove forests. Still, most importantly, an alternative and a cheap source of cooking must be provided to the low-income earners in the community to reduce the incidence of deforestation and reduce the dependence on the local community for the use of mangrove wood as a primary source of energy. **Figure 8** shows wood harvesters in Goi, a small Ogoni community, in boats returning to their base after harvesting mangroves.

Population growth, urbanisation, and construction are significant threats to the mangrove ecosystem in the Niger Delta. As the population of the people of the Niger Delta increases, the demand for forest products such as wood also increases. Population growth would automatically mean people require more land for farming, building, and other social needs and, by extension, additional pressure on the surrounding forest. Construction of roads and oil and gas facilities, such as pipeline networks, have extensively degraded mangrove forests in the Niger Delta. Proper enlightenment of the local community on the need to protect the forest must be a task that the government and civil societies would have to embark on so mangrove forest goods and services can be used sustainably in the Niger Delta. Despite the threat discussed so far, the effect of petroleum hydrocarbon spillage and the invasion of Nipa palm in mangrove habitat in the Niger Delta remain two of the biggest threats currently facing the Niger Delta mangrove forest at the moment and will be discussed broadly.

Figure 8.
Shows wood harvesters in Goi Gokana LGA, a small Ogoni community, in boats returning to their base after harvesting mangroves.

4.1 Nipa palm invasion of mangrove forest in the Niger Delta

Nypa fruticans, also known as Nypa or Nipa palm, are commonly found in Asia. They are large green palms forming loose clumps from a subterranean stem. Nipa palm has clumps of individual erect and large leaves that can grow as long as 6 m. In Asia, Nipa palm seeds and saps are edible and used for thatching. Nipa palms are not natural flora in the Niger Delta mangrove forest; their presence in the Niger Delta was due to their introduction in Calabar, Nigeria, in 1906 from a Singapore botanical garden. In 1964, thousands of seeds originating from Malaya were planted throughout the brackish swamps of the Niger Delta. These were the major points where Nipa was introduced to the Niger Delta. Introducing these plants in the Niger Delta provided an alternative palm for the people to the native palm (*Elaeis guineensis*), mainly used for thatching, food, and alcoholic wine [13]. It is safe to say that at the time these decisions were made, the value of mangroves had not been fully understood by those who made the decision, and the unintended consequences of Nipa palm invading the mangrove forest were not anticipated.

In the Niger Delta, Nipa palm has become an invasive species and is taking over mangrove habitat. Nipa palm can be found along the coastline of the Niger Delta mangrove forest down to Lagos and Ogun. Unfortunately, the local community does not use Nipa palm in the Niger Delta for any purpose, as they are not the palm of choice for thatching; their sap and seeds are edible in other parts of the world but are currently not eaten in Nigeria. Nipa palm grows on the fringes of the Niger Delta mangrove platform, competing with mangroves for space, food, and habitat, and expanding its reach when the opportunity provides itself. In a healthy, matured mangrove forest, the only available areas Nipa can grow are the fringes and sometimes channel edges of the mangrove platform and along the coastline.

The current degradation of mangrove forests in the Niger Delta has provided Nipa palms with the opportunity to thrive in areas where mangroves would naturally outcompete them in the Niger Delta. This has allowed Nipa palm to find its way into the heart of mangrove platforms and establish itself, thereby colonising areas where mangrove plants should grow. Nipa palm and mangroves have similar modes of seed dispersal; unfortunately, Nipa tends to outgrow and outcompete mangroves when the two plants are simultaneously established on the same soil in a mangrove habitat [14]. Nipa palm invasion of mangroves has been described as one of the significant threats to the mangrove ecosystem in the Niger Delta, the growth and spread of Nipa palm, if not checked, would spiral out of control. The current effort to stop the spread of the plant in the region has been to mechanically remove the plant from the mangrove habitat by cutting down the plant and taking their seeds away, as it is currently done in Bodo, a small Ogoni Community where restoration efforts are being made in the Niger Delta [7].

The sustainability of mechanical removal of Nipa palm would require a concerted effort of the entire Niger Delta and Nigeria, including its west African Neighbours whose coastlines have also been overrun by Nipa palm. Today Nipa palm has become not just an invasive species in the Niger Delta; it has also formed part of the associated mangrove species in the Niger Delta since their number cannot be overlooked and can no longer be considered an exotic species if nothing is done about their rapid growth and spread. **Figure 9** shows the Nipa palm growing on the fringes of a mangrove platform in the Niger Delta and the mechanical removal of the Nipa palm from a remediated mangrove forest in the Niger Delta.

Figure 9.
Shows the Nipa palm growing on the fringes of a mangrove platform in the Niger Delta and the mechanical removal of the Nipa palm from a remediated mangrove forest in the Niger Delta.

4.2 Crude oil spills on mangroves in the Niger Delta

Petroleum hydrocarbon is one of the vast natural resources in the Niger Delta and the primary income of Nigeria today. Unfortunately, the Niger Delta of Nigeria has been on the receiving end of many environmental disasters resulting from petroleum hydrocarbon pollution. In the Niger Delta, some oil and gas exploration well heads are located on mangrove forests, and vast networks of oil and gas pipelines in the Niger Delta run through mangroves forest or creeks and estuaries adjourning mangrove forests; this has made it possible for mangroves to be readily degraded by petroleum hydrocarbon whenever there is an accident, sabotage or equipment failure on any of these facilities. The oil spill history in the Niger Delta dates back to when commercial exploration of oil began in Oloibiri, current Bayelsa state, in February 1958. Since the commercial exploration of oil and gas in Nigeria, thousands of oil spill incidents have been reported in the Niger Delta. Some of these have been noted to affect mangrove habitat; one such incident where large-scale mangrove destruction occurred in the Niger Delta is in Ogoni and Bonny in River state, Nembe in Bayelsa state, and many other regions in the Niger Delta.

A case study of a significant oil spill incident in the Niger Delta happened in 2008 and 2009 along the Trans Niger Pipeline (TNP), which carries Bonny light crude oil to an export terminal in Bonny River State. The TNP is critical to Nigeria's crude oil export and is jointly operated by the Nigerian government and a major international oil and gas company; the TNP carries 180,000 barrels daily to the Bonny export terminal. In October 2008, there was a leak along the TNP, which happened to be along the creeks in a community known as Bodo. After a joint investigation, it was discovered that the leak was caused by equipment failure due to corrosion due to the old pipeline conveying crude, which was laid in the 1960s. Locals reported that the leaks in the pipeline occurred for weeks before it was clamped, and thousands of barrels of crude were spilled into the surrounding creeks, which eventually damaged hectares of mangrove forest along Bodo, Goi, Kpor, and Bomu in Gokana River state. In February 2009, another leak occurred in Bodo along the TNP, a few kilometres away from the previous leak; no data was put out on the exact amount of oil introduced to the environment. The effect of the two spills resulted in the degradation of over 1000 hectares of mangrove forest and the destruction of fauna and flora within and around the affected areas [15, 16].

4.3 The effect of crude oil spill on mangroves

Mangroves are susceptible to oil exposure depending on the degree of oiling, length of exposure, and, to an extent, the type of crude oil. Mangroves exposed to crude oil in sufficient amounts to cause harm will suffer from their leaves turning yellow and falling off (defoliation) and eventually death. The smaller and younger trees or plants whose stems and leaves are entirely coated by oil are usually among the first to die off because the plant's lenticels and leaves used for transpiration and salt excretion are covered with crude, thereby preventing the natural biological function of the entire plant [17].

Following the two spills in Bodo creek, crude oil floated freely on the surface of the creek; being an intertidal area, crude oil was constantly being carried into intertidal platforms where they are deposited on sediments, leaves, stems, and prop roots of mangroves. A few weeks after the spills, all plants that were coated with oil from their stems to their leaves died; it did not take long after areas that had high crude oil deposits, all plants gradually started dying off, and hectares of mangrove forest was left with stumps of interwoven dead mangrove plant. Years after the spill, the entire dead plant stumps collapsed, leaving an open island of degraded mangrove platform with dead mangrove stumps littered all over the platform.

The effect of crude oil on mangrove habitat has a ripple effect on the faunal community and the livelihood of the people who depends on mangroves for the many goods and services it provides to the locals in the community. Most Niger Delta communities, as noted earlier, are heavily dependent on fishing, and polluted creeks resulting from crude oil spills are often devoid of fish even after cleanup, it takes a significant amount of time for the fish communities in most of these creeks to be back to its pristine conditions. The cascading effect of spill damage on mangroves forest has been experienced in many Niger Delta communities in Ogoni Rivers State; due to the death of mangroves closer to the shorelines where the people reside, wood harvesters have had to go deeper into areas where the spill barely affected to harvest wood, this has put enormous pressure on the forest as trees which are not matured enough are harvested leading to rapid deforestation of the mangrove forest. Furthermore, the spill's effect on invertebrate and fish communities has put enormous pressure on the creeks due to the trawling of small and immature fish due to the depleted fish stock in the creeks. Juvenile mangrove plants have also been harvested in these areas, reducing the chances of natural recovery. As a result, human-mediated revegetation is required to restore the lost ecosystem.

Although the damage from a large-scale oil spill has acute and chronic implications, pollution from artisanal (Illegal) refining, bunkering, and crude oil theft have also destroyed hectares of mangrove forest in the Niger Delta, and most of these destructions are largely undocumented. Artisanal or illegal refining of crude oil in the Niger Delta, locally known as "Kpo-Fire," is the small-scale processing of stolen crude by heating it at high temperatures to distill it into its fractional components. The mangrove forests and other parts of the Niger Delta have become spots and makeshift refineries for artisanal refining crude oil. The implication of this is that mangroves are cleared to set up a refinery, and in most cases, mangrove trees are harvested to provide energy for heating the crude at very high temperatures; the effluent, tar, and asphalt, which are usually bi-product of the refined crude are often discharged in pits in or around the mangrove forest. Crude oil used for artisanal refining in the Niger Delta is stolen chiefly from pipelines running through the mangrove forest. The point at which the pipes are compromised, if not well-engineered, is usually a source of spill in the mangrove forest; transportation of the crude and storage of the crude, including refined artisanal products, are usually significant spill sources in the Niger Delta. Artisanal refining camps in the Niger Delta have been on the increase in the Niger Delta despite several government clamps down; the more camps created, the more pressure on mangroves somewhere in the Niger Delta; unfortunately, most of the damage from artisanal refining and crude oil theft on mangrove forest is not documented as such the extent of damage caused by artisanal refining on mangrove forest in the Niger Delta today is difficult to quantify [10].

Despite the damage from the crude oil spill on mangroves in the Niger Delta, efforts are currently ongoing to restore hectares of damaged mangrove forest in the Niger Delta, mainly due to the UNEP report [6] and a few litigation cases between the host community and international oil and gas companies responsible for the spill in partnership with Nigerian government regulatory agencies. Some of these efforts have shown great restorative potential, and remediation, cleanup, and revegetation have begun in earnest.

5. Mangrove restoration in the Niger Delta

In the Niger Delta, mangrove restoration has been done mainly on a small scale. A mediation-driven mangrove restoration program is currently Nigeria's most extensive [18]. Although the government-backed cleanup of Ogoniland would involve mangrove restoration, the remediation phase of the cleanup of intertidal areas is currently ongoing. Mangrove forest impacted significantly by crude oil spills requires cleanup/remediation of the contaminated mangrove sediments, planting, and monitoring of mangroves. Mangrove ecosystems are intertidal environments that are usually very challenging to clean up or remediate after a spill. For the first time in the Niger Delta, the shoreline cleanup assessment technique (SCAT) has been deployed to clean up/remediate contaminated mangrove sediment in the Niger Delta [7, 16].

5.1 The use of SCAT in the Niger Delta

The Exxon Valdez spill in 1989 birthed the origin of SCAT, and since then, the technique has been used in several spill cleanups globally. SCAT was first used in the Niger Delta in 2019 and is currently used to clean up spill sites in the Niger Delta. SCAT involves a series of surveys before, during, and after the cleanup/remediation operation. SCAT rapid surveys are initial surveys done during the SCAT process to determine the level of contamination and areas affected by the spill. The SCAT rapid survey serves as a baseline where blueprints for remedial/cleanup actions are designed. SCAT assessment surveys are done to delineate work areas and get in-depth information to characterise work areas. SCAT confirmation or verification surveys are done after cleanup/remedial actions have been done in the affected areas [19].

SCAT process requires the participation of all stakeholders in the remediation process; stakeholders in the Niger Delta usually include the host communities, the oil and gas company, government regulatory agencies, a mediation body (if set up), and civil society organisations. Representatives of all stakeholders would be available at all stages of the SCAT process. Cleanup or remediation is given to contractors with expertise in oil spill remediation. The stakeholders often agree upon remediation techniques, and the SCAT team confirms remediation work. The SCAT team is composed of a representative of all stakeholders at every site or work area to be confirmed; SCAT confirmation is done by digging three pits and assessing the level of oiling in the pit. SCAT assessors are usually trained in SCAT techniques to assess the pit; remediated site maps are designed and produced by the management and SCAT team lead, and coordinates of the sites are handed over to the SCAT team for verification and confirmation.

Pit oiling assessment is critical to verifying and confirming remediated sites; in the cleanup process at Bodo, pit oiling assessment for confirmation of a site is a maximum of 25% oiling per pit. A work area can have as many sites as possible, depending on the size of the area. Irrespective of the number of sites in the remediated area, each site must have three pits, usually in a trisection, with the sum of pit oiling in all three pits less than 75%. Although this process is subjective and open to bias, the oiling level for each pit is agreed upon by the entire stakeholders in the SCAT team. A standard pit oiling chart also guides the SCAT team in their assessment when determining oiling levels in the pit.

5.2 Mangrove planting and monitoring

After cleanup/remediation is completed, the next phase of the restoration program is planting mangroves. In the Niger Delta, mangroves are planted as propagules, sprouted propagules, or seedlings. While propagules are seeds from mangrove plants (**Figures 1–4**) harvested and put directly on restoration sites to grow, sprouted propagules and seedlings are grown or nursed in a nursery before being transplanted to their restoration site after a certain period. Seedlings are preferred as the life stage of choice in the Niger Delta for restoration, as observed in the Bodo revegetation projects. Seedling also increases restoration costs but improves the survivability of the overall plants during restoration. Seedlings will require the setting up nurseries, usually found in some Niger Delta communities close to intertidal areas. Nurseries are sited close to intertidal areas to avoid the constant need to water the plant with brackish water when the nursery is located far away from intertidal areas [20].

Most mangrove nurseries in the Niger Delta have only *Rhizophora spp.* Seedlings grown in them might be unconnected with the fact that *Rhizophora spp.* is the dominant mangrove species in the Niger Delta, and most locals in the community do not see other mangrove species as part of the mangrove community in the Niger Delta. Mangrove planting involves transporting mangrove seedlings to areas where they would be transplanted; care must be taken during transportation since some mangrove areas might be inaccessible on foot. Mangrove seedlings should be planted in a single bag, although to save cost, specific mangrove nurseries in the Niger Delta plant multiple propagules in a single bag. Single bags are advised to reduce stress on the plant during transplanting and to ensure that the soil and the plant are placed in the ground together. **Figure 10** shows a mangrove nursery in the Niger Delta and how propagules are planted in single and multiple propagules per bag.

Monitoring mangroves is done to ensure their growth and survivability and to identify areas where significant losses occur, the cause(s) of the loss, and the need to replant after identifying why the loss occurred. Mangrove monitoring is critical to the survivability and growth of large-scale mangrove revegetation programs. The monitoring program in Bodo involves collecting specific data from marked plants in different planting areas. Monitoring is done at intervals of one, six, 12, and 18 months; specific data such as the increase in height, number of leaves and branches, oiling conditions number of prop roots or pneumatophores are collected at every monitoring interval. An overall count of the number of dead mangroves is done after the first

Figure 10.
Is a picture of a mangrove nursery in the Niger Delta and how propagules are planted in single and multiple propagules per bag.

month of planting, where deaths exceed 5%; in a delineated planting area, the dead plants are replaced, and monitoring is scheduled for another month after replacing the dead plants. The process is repeated till the death rate in a particular planting area is less than 5%. Restoration of mangrove habitat is a long-term project as mangroves take years to fully mature.

5.3 Challenges of mangrove restoration in the Niger Delta

Corruption, weak government regulatory system, and lack of environmental justice have been the bane of mangrove forest restoration in the Niger Delta. Oil spills in mangrove ecosystems remain un-remediated, and where remediation Is done, they need to be done correctly. Also, most Niger Delta communities impede the cleanup process by demanding outrageous amounts from the oil and gas companies even after compensation has been paid out; in some cases, community leaders demand money to be used for the cleanup process be paid to them.

6. Conclusion

Mangroves of the Niger Delta are the largest and one of the most threatened mangrove ecosystems in Africa. The Niger Delta mangrove is a very significant ecosystem in the Niger Delta as they provide a variety of environmental, economic, and social goods and services to the people of the Niger Delta. Unfortunately, the importance of this vital ecosystem is not well known by most people in the Niger Delta; as such, the Niger Delta mangroves are depleted in certain areas at an alarming rate. Significant threats to the Niger Delta mangroves include Nipa palm invasion, oil spill pollution, deforestation, and urbanisation. Mangroves in the Niger Delta must be protected from the threats it is currently facing for the sustainability of this vital ecosystem. It is also essential to educate the communities around mangrove areas on the importance of protecting these forests; the government must, as a matter of national interest, map out areas of the Niger Delta mangrove forest for conservation; this would ensure the sustainability of the mangroves.

Conservation is a long-term strategy for protecting the Niger Delta mangroves; in the interim, both national and state governments must collaborate with all state actors in the oil and gas industry in the region to ensure the complete cessation of spills in mangroves and other parts of the Niger Delta, and as a matter of urgency stop illegal bunkering and artisanal refining of crude in the Niger Delta. A policy framework must also be drawn up on how to control the rapidly spreading Nipa palm currently invading mangrove forest in the region. Despite the significant threats affecting the mangroves in the Niger Delta, the sustainability of the region's mangroves is still very realistic if current threats are addressed.

Acknowledgements

I want to acknowledge Dr. Erich Gundlach of the Bodo Mediation Initiative, under whose platform I have been able to learn about the mangroves of the Niger Delta. I also appreciate his mentorship skills and willingness to impart knowledge. Special thanks to my colleagues Nicholas Story, Peter Lenu, and Bariton Lezabby for contributing to various aspects of this book chapter.

Author details

Izuchukwu Uche
Centre for Environmental Management and Control, University of Nigeria, Enugu, Nigeria

*Address all correspondence to: izucals@gmail.com

References

[1] Global Mangrove Watch Nigeria Mangrove Net Change. 2022. [Internet] Available from: globalmangrovewatch.org. [Accessed: 2022-10-23]

[2] Numbere AO. Camilo GR Mangrove leaf herbivory along a hydrocarbon pollution gradient in a mangrove forest (*Rhizophora racemosa*) in the Niger River Delta, Nigeria. Journal of Petroleum & Environmental Biotechnology. 2018;**10**:391. DOI: 10.35248/2157-7463.19.10.391

[3] Zabbey N, Tanee BG. A Training Manual on Mangrove Restoration in Coastal Communities of the Niger Delta Nigeria. CEHRD, Port Harcourt Nigeria: Mobility Concept Press; 2021. p. 38

[4] Gundlach ER. Oil-related mangrove loss east of Bonny River, Nigeria. In: Makowski C, Finkl CW, editors. Coastal Research Library (CRL): Threats to Mangrove Forests: Hazards, Vulnerability, and Management Solutions. Coastal Research Library. Switzerland, Cham: Springer; 2018. pp. 267-321. DOI: 10.1007/978-3-319-73016-5_13

[5] Numbere AO. Mangrove habitat loss and the need for the establishment of conservation and protected areas in the Niger Delta, Nigeria. In: Muscarella CM, Ortiz AC, Canas RQ, editors. Habitats of the World – Biodiversity, and Threats. London: IntechOpen. DOI: 10.5772/intechopen.89623

[6] UNEP. Environmental assessment of Ogoniland, United Nations Environment Programme, Nairobi, 257. 2011. Available from: http://www.unep.org/nigeria. [Accessed: 23/10/2022]

[7] Gundlach ER, Iroakasi O, Story NI, Nwabueze V. Bodo Project, Ogoniland, Eastern Niger Delta. Part 2 – Innovative close-out procedures for oil-contaminated mangrove habitats. International Oil Spill Science Conference, Halifax, Canada, 21-25 March 2022. 2022

[8] Iroakasi O, Gundlach ER, Devaull G, Nwabueze V, Bonte M. Bodo Project, Ogoniland, Eastern Niger Delta. Part 1 – Application of RBCA and NEBA to determine site-specific target levels. International Oil Spill Science Conference 2022 March 21-25, Halifax, Canada. 2022

[9] Asanebi DH. A concise view of Niger Delta region of Nigeria: An interpretation of a Nigeria historian. International Research Journal of Interdisciplinary & Multidisciplinary Studies (IRJIMS). 2016;**2**(10):56-63

[10] Efenakpo O, Chris D, Onuchukwu N, Kejeh A. Illegal crude oil refining and its implications on the Niger Delta's ecosystem. DOI: 10.13140/RG.2.2.23863.39840

[11] Ikezam P, Elenwo EI, Oyegun CU. Effects of artisanal refinery on the environment, public health and socio-economic development of communities in the Niger Delta region. Environmental Management and Sustainable Development, Macrothink Institute. 2021;**10**(3):97-111. Available from: https://ideas.repec.org/a/mth/emsd88/v10y2021i3p97-111.html

[12] Gundlach ER, Giadom FD, Akpokodje EG, Bonte M, Tse AC, Ekeocha NE, et al. Core sediments and oil chemistry from contaminated mangroves in eastern Niger Delta, Ogoniland, Nigeria. Marine Pollution Bulletin. 2021:171 112714. DOI: 10.1016/j.marpolbul.2021.112714

[13] Isebor CE, Ajayi TO, Anyanwu A. The incidence of Nipa Fruticans and its impact on fisheries production in the Niger Delta Mangrove ecosystem. 16th Annual Conference of the Fisheries Society of Nigeria (FISON) 4-9 November 2001 Maiduguri, Nigeria. 2011. pp. 13-16. Available from: http://hdl.handle.net/1834/21300

[14] Numbere AO. Effect of soil types on growth, survival, and abundance of mangrove (*Rhizophora racemosa*) and Nypa Palm (*Nipa fruticans*) Seedlings in the Niger Delta, Nigeria. American Journal of Environmental Sciences. 2019;**15**(2):55-63. DOI: 10.3844/ajessp.2019

[15] Pegg S, Zabbey N. Oil and water: The Bodo spills and the destruction of traditional livelihood structures in the Niger Delta. Community Development Journal. 2013;**48**(3):391-405. DOI: 10.1093/cdj/bst021

[16] Gundlach ER, McArthur A, Iroakasi O, Bonte M, Giadom FD, Shekwolo P, et al. Cleanup and restoration of 1000-ha of oiled Mangroves, Bodo, Eastern Niger Delta, Nigeria. IOSC Proceedings. 2021;**1**:1-25. DOI: 10.7901/2169-3358-2021.1.688932

[17] Wilson M, Hale C, Maung-Douglass E, Partyka M, Sempier S, Skelton T, et al. Impacts of Oil on Mangroves. Oil Spill Science: Sea Grant Programs of the Gulf of Mexico; 2019. (GOMSG-G-19-010)

[18] Gundlach ER, Bonte M, Story NI, Iroakasi O. Using high-resolution imagery from 2013 and 2020 to establish baseline vegetation in oil-damaged mangrove habitat before large-scale post-remediation planting in Bodo, Eastern Niger Delta, Nigeria. Remote sensing applications: Society and Environment. 2022;**28**:100831. DOI: 10.1016/j.rsase.2022.100831

[19] Environment and Climate Change Canada (ECCC). Shoreline cleanup assessment technique (SCAT) manual. In: Triox Environmental Emergencies, Owens Coastal Consultants. 3rd ed. Ottawa: Environmental Mapping Ltd; 2018. p. 40

[20] Zabbey N, Tanee F. Assessment of Asymmetric Mangrove Restoration Trials in Ogoniland, Niger Delta, Nigeria: Lessons for Future Intervention. Ecological Restoration. 2016;**34**(2):245-257. DOI: 0.3368/er.34.3.245

Chapter 2

The Niger Delta Mangrove Ecosystem and Its Conservation Challenges

Anthony E. Ogbeibu and Blessing J. Oribhabor

Abstract

The Niger Delta region of Nigeria is a biodiversity hotspot, rich in fauna and flora, with robust ecological processes that drive the economy of the local communities. However, the area has not been formally recognized as a global biodiversity hotspot by international agencies such as the International Union for the Conservation of Nature (IUCN). The Niger Delta abounds in several species of mangroves, but the most dominant ones are the red (*Rhizophora racemosa*), black (*Laguncularia racemosa*), and white (*Avicennia germinans*) mangroves. The mangrove ecosystem in the Niger Delta is fast being replaced by Nipa palm (*Nypa fruticans*), yet no result-oriented action plan has been put in place to conserve the mangrove. The serious menace of oil pollution and overexploitation of fisheries and mangrove resources, in addition to environmental pressures, as well as poor recognition of brackish water bodies and aquaculture, causes serious danger to the sustainability of Nigeria's brackish water and the various fish resources, wildlife, and, inevitably, the coastal communities of the nation. This chapter provides essential information on the rich Niger Delta mangrove ecosystem and current conservation challenges and recommends strategies for sustainable management.

Keywords: Niger Delta, mangroves, biodiversity, water quality, conservation

1. Introduction

Nigeria is the most populous country in Africa. Since the attainment of independence in 1960, it has continuously grown in both size and influence, with an approximate population of 203 million people, accounting for 47% of the total West African population [1]. It hosts a variety of terrain and climate. The seven main ecological zones are the Mangrove Swamp and Coastal Vegetation, Freshwater Swamp Forest, Lowland Rain Forest, Derived Savanna, Guinea Savanna, Sudan Savanna, and Sahel Savanna [2]. The Niger Delta geographic basin is defined by the many tributaries of the lower River Niger, as it enters the Atlantic Ocean through its many tributaries and estuaries. It is a region with very high biodiversity as well as a dense human population. At the base of the Niger Delta, formed by a network of many creeks, is Africa's largest mangrove forest, the third largest in the world.

The Niger Delta is located on the Atlantic coast of Southern Nigeria, which extends between latitudes 4° 2′ and 6° 2′ north of the equator and is 50 2′ east of the Greenwich meridian [3, 4] and bordered by the Atlantic Ocean in the south, and Cameroon in the East. It occupies a surface area of about 112,110 km^2. It represents about 12% of Nigeria's total surface area and is almost two-thirds of the entire population of Nigeria (i.e., 200 million). The region is made up of nine of Nigeria's constituent states (**Table 1**) [5]. The region has a yearly rainfall of 3000–4500 mm and an average temperature of 27°C. The wet season is relatively long, occurring from March to October, and the dry season from November to February [6, 7].

The Niger Delta floodplain occupies 7.5% of Nigeria's total land mass and is her largest wetland. It is the third largest drainage basin in Africa. The Delta's environment includes four ecological zones: coastal barrier islands, mangrove swamp forests, freshwater swamps, and lowland rainforests. This vastly well-endowed ecosystem contains one of the Earth's highest concentrations of biodiversity. Apart from supporting abundant flora and fauna, arable terrain that sustains a wide variety of crops, lumber, or agricultural trees, it also harbors more freshwater fish species than any other ecosystem in West Africa.

The biodiversity of the Niger Delta is of regional and global significance. Barrier islands, estuaries, mangroves, creeks, and freshwater swamps characterize the Niger Delta ecosystem. The Niger Delta region, considered a "*Biodiversity Hotspot*" by biodiversity experts, abounds in many locally and globally endangered species with robust ecological processes that drive the economy of the local communities.

Despite the importance of forest resources, there has been a rapid decrease recently due to the demand from an ever-increasing population. Drivers of forest loss in the different forest types are complex, including illegal logging from commercial loggers, due to high timber demand and communal logging of firewood for domestic cooking [8]. Mangrove ecosystems in the Niger Delta are fast being replaced by Nipa palm (*Nypa fruticans*). Yet no result-oriented action plan has been put in place to conserve the dwindling mangrove. The serious menace of oil pollution and overexploitation of fisheries and mangrove resources, in addition to other environmental pressures, as well as poor interest and recognition of brackish water bodies and aquaculture, cause serious danger to the sustainability of Nigeria's brackish water and

State	Land area (km^2)	Population	Capital
Abia	4,877	5,106,000	Umuahia
Akwa Ibom	6,806	5,285,000	Uyo
Bayelsa	1,107	2,703,000	Bayelsa
Cross River	21,930	4,325,000	Calabar
Delta	17,163	5,681,000	Asaba
Edo	19,698	4,871,000	Benin City
Imo	5,165	5,283,000	Owerri
Ondo	15,086	4,782,000	Akure
Rivers	10,378	7,679,000	Port Harcourt
Total	**112,110**	**45,715,000**	

Table 1.
Niger Delta states of Nigeria, their land areas, population, and capitals [5].

the various fish resources, wildlife, and, inevitably, the coastal communities of the nation [9].

The importance of mangroves as nurseries has been one of the reasons to support their conservation and management [10]. Many African, Latin American, and Asian countries are now estimated to have lost at least 50% of their original mangrove area [11]. The current wave of global concern for the maintenance of essential ecological processes and life support systems, preservation of genetic diversity, and the sustainable utilization of species and ecosystems has led to the intensive survey of nature reserves for their ecological descriptions, biodiversity, and conservation status [12]. According to the International Union for the Conservation of Nature and Natural Resources (IUCN) report [13] on the global status of mangrove ecosystems, 18 countries have established mangrove reserves to safeguard the habitat and associated species. The mangrove zone of the Niger Delta is said to traverse parallel to the coast and reaches between 15 and 45 km inland. This deep belt of mangrove forest protects the freshwater wetlands in the Inner Delta [14].

The Niger Delta mangrove ecosystem, a fragile ecosystem increasingly being threatened by Nipa palm (*Nypa fruticans*) encroachment and industrial/oil pollution, is presently receiving urgent national and international attention for sustainable management and conservation. The problem is that the area has not been formally recognized as a global biodiversity hotspot by international agencies such as the IUCN. This Book Chapter provides essential information on the rich Niger Delta mangrove ecosystem and its conservation challenges and recommends strategies for sustainable management.

2. Terrestrial ecosystems of Nigeria and mangrove ecosystems of the Niger Delta

Nigeria has many regions with rich biodiversity serving as home to critical ecosystems. Among these regions are the Cross River basin rainforest, mountains along the Cameroon border with Nigeria, and the Niger Delta coastal creeks (**Figure 1**) [15, 16]. The Niger Delta mangrove is Africa's largest mangrove ecosystem and the world's third largest after India and Indonesia, the third largest drainage basin in Africa, and Africa's largest river delta [17, 18].

The coastal region of Nigeria extends from Benin Republic/Nigeria border in the west to the Cameroun/Nigeria border in the east, a distance of 800 km but with a coastline of 853 km [19]. This coastline has interfered with a series of estuaries that empty into the extensive lagoon system in Lagos and Ondo State. At least twenty-two (22) estuaries exist between the Benin River in Delta State coastal region and the Cross River in Akwa Ibom State. In the Delta and River State areas, the estuaries join with tributaries and distributaries of the lower Niger River/Benue system to form an anastomosing system of fresh and brackish water creeks, backwaters, and flood plains in what is better described as the Niger Delta basin swamps [20]. The creeks, floodplains, lagoons, and rivers in Lagos State account for approximately 22% of the 790 km^2 land mass [21].

The brackish water sector consists of estuaries, beach ridges, intertidal mangrove swamps, intersecting rivers, and winding saline creeks. The exact area of the brackish water for the entire Nigerian coastline is unknown except for the Niger Delta portion, the area of which has been estimated [22] and includes the constant saline creeks that form an area of about 1000 km^2 and the intertidal mangrove swamps consisting an

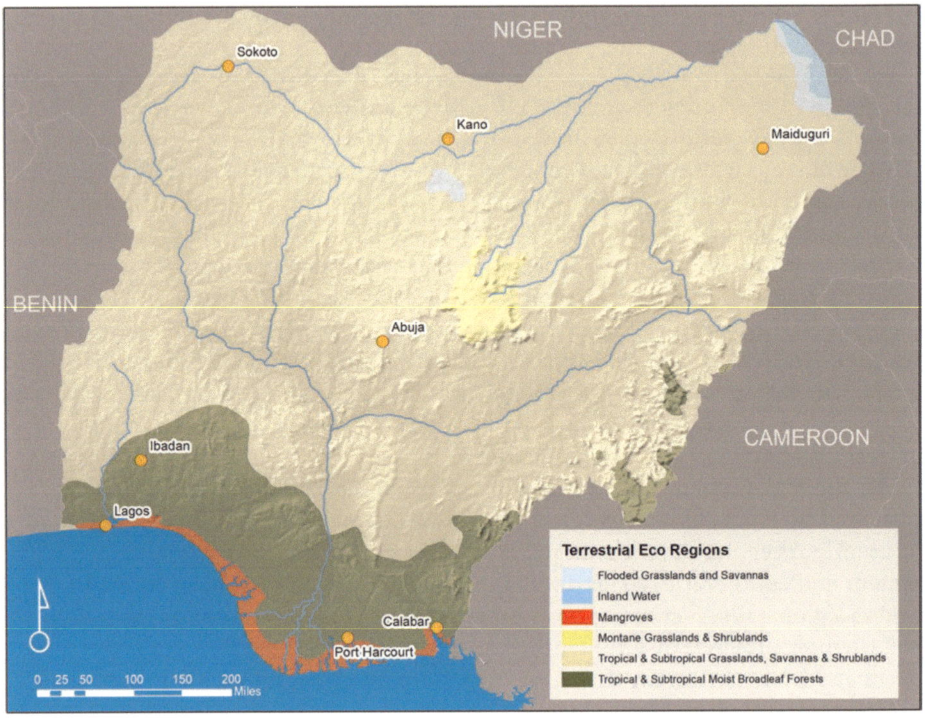

Figure 1.
Nigeria's terrestrial ecoregions [15].

area of about 5048 km^2. In addition, there are intersecting rivers and estuaries plus beach ridges with an area of about 679 km^2 and 688 km^2, respectively. The total brackish water area of the Niger Delta is therefore approximately 7415 km^2. The inland area of the mangrove swamp fairly delimits the brackish water area. The brackish water region of Nigeria's coastal zone between the Benin River and Cross River has been adequately delineated [23].

Mangrove forests could be cleared for aquaculture ponds, salt pans, agricultural use, including rice fields, airport and road construction, port and industrial development, resettlement, and village development [24]. They are, however, important factors in stabilizing the shoreline (http://www.panda.org). Some endangered and potentially vulnerable species are endemic in this area. The mangrove swamps and adjoining freshwater swamps in the Niger Delta provide critical habitats for many endangered animal species, important wildlife, and migratory and resident aquatic reptiles, birds, and mammals. These include crocodiles, Nile monitor lizards, parrots, duikers and antelopes, monkeys, bush pigs, bush cow or buffalos, sitatungas, hippopotamus, diving pelicans, and deep swimming shags [25]. The subtidal prop root of mangrove habitats serves as nurseries for fishes of economic importance [26]. The interest in protecting mangroves worldwide is due to their purported importance to fisheries and a number of rare and endangered species [27]. However, because the same mangrove species can often occur under marine, estuarine, and freshwater conditions, a wide variety of fish assemblages can be found among their inundated "mangrove habitats." As such, mangrove habitats likely play a variety of roles in the lives of associated fishes, feeding areas for some species or life stages, daytime refugia for others, and nursery and/or nesting areas for yet more. Mangroves of the Atlantic coast of Africa, including the Niger Delta, on account of their gentle gradient of sediment, are sensitive/fragile [28]. The rich biodiversity and high aquatic productivity of the West African coastal waters are constantly undergoing bio-modification due to the adverse impact of human activities [29].

2.1 Mangrove flora of the Niger Delta

The most symbolic mangrove species in Nigeria consists of six species in three families, namely: Rhizophoraceae (*Rhizophora racemosa, R. harrisonii, and R. mangle*), Avicenniaceae (*Avicennia africana*), and Combretaceae (*Laguncularia racemosa* and *Conocarpus erectus*). In terms of biodiversity, the Niger Delta mangrove ecosystem is one of the richest wetlands globally by researchers.

There are several species of mangroves in the Niger Delta, with the most dominant ones being red (*Rhizophora racemosa*), black (*Laguncularia racemosa*), and white (*Avicennia germinans*) mangroves. They provide highly productive habitats and ecological niches for reptiles, monkeys, birds, fishes, shrimps, mollusks, and other wildlife species, and a number of other ecological services [30]. Also prominent but less studied and uncommon around core mangrove forests are Button wood mangroves (*Conocarpus erectus*). They are common in inland sandy soil areas [5]. The following eight true West African mangrove species are found in Nigeria, particularly in the Niger Delta region: *Avicennia germinans*, *Rhizophora mangle*, *Rhizophora racemosa*, *Rhizophora harrisonii*, *Laguncularia racemosa*, *Conocarpus erectus*, *Nypa fruticans,* and *Acrostichum aureum*. The Families Rhizoporaceae, Avicenniaceae, and Combretaceae, such as *Rhizophora racemosa*, *Rhizophora harrisonii*, *Rhizophora mangle*, *Avicennia africana* (White mangrove), *Laguncularia racemosa*, and *Conocarpus erectus* and a few other stunted shrubs and woody species are the most dominant plant species.

Most locations in the Niger Delta have similar mangrove species composition [5]. The floristic composition of the plants and family distribution in the Mangrove Swamp Forest of Anantigha in Calabar South Local Government Area (LGA), Cross River State, Nigeria, is shown in **Table 2** [30]. The plant population consisted of *Rhizophora racemosa*, *Nypa fruticans*, *Avicennia germinans*, *Acrostichum aureum*, *Drepanocarpus lunatus*, *Laguncularia racemosa*, *Hibiscus tiliaceus*, *Thespesia populnea*, *Pandanus candelabrum*, *Carapa procera*, *Raphia hookeri*, *Conocarpus erectus,* and *Chrysobalanus spp.* [30]. The most abundant species in the area are *Rhizophora racemosa*, *Nypa fruticans*, *Avicennia germinans,* and *Acrostichum aureum*. *Rhizophora racemosa* had the highest species importance value (SIV), followed by *Nypa fruticans*, an exotic palm that encroaches on the mangrove and dominates the indigenous oil palm, *Elaeis guineensis* [30]. For the grass species, *Dalbergia ecastaphyllum* is reported to have the highest diversity (**Table 3**). Species distribution from seaward to landward areas indicates that core mangrove species were found on the seaward side, whereas the non-mangrove species were found in the landward direction [5].

A related study documented the common plant species in Tunu and Kanbo in South Forcados of the Niger Delta ecosystem. The predominant vegetation type within the Tunu/Kanbo field is the mangrove swamp forest type, attested to by the floral composition (**Table 3**) [31].

The dominant plant with the highest frequency of occurrence is the dwarf red mangrove, *Rhizophora mangle*. The tall red mangrove *R. racemosa* was confined to the edge of the creeks and rivers. Herbaceous grasses, fern, and sedges such as *Acrostichum aureum, Paspalum vaginatum,* and *Fimbristylis spp.* occurred on the forest floor below the mangrove canopy. Epiphytes were abundant on the tree trunks. The

S/N	Plant Species	Families	Species density (Individuals/ Ha)	Siv
1	*Rhizophora racemosa*	Rhizophoraceae	45	34.892
2	*Nypa fruticans*	Arecaceae	23	19.245
3	*Avicennia germinans*	Avicenniaceae	18	15.468
4	*Acrostichum aureum*	Pteridaceae	15	13.309
5	*Drepanocarpus lanatus*	Fabaceae	10	9.892
6	*Laguncularia racemosa*	Combretaceae	8	8.273
7	*Hibiscus tiliaceus*	Malvaceae	7	8.094
8	*Thespesia populnea*	Malvaceae	5	6.295
9	*Pandanus candelabrum*	Pandanaceae	4	5.935
10	*Carapa procera*	Meliaceae	3	3.957
11	*Raphia hookeri*	Palmaceae	2	3.777
12	*Conocarpus erectus*	Combretaceae	1	1.799
13	*Chrysobalanus spp.*	Chrysobalanaceae	1	1.439

Table 2.
Plants found in Anantigha mangrove swamp forest with their species density, relative density, and species importance value (SIV) [30].

commonest species include *Bulbophyllum oreonastes* and the moss *Ocoblepharum sp.* The mangroves are admirably adapted to the brackish environment. They have stilt roots that are covered with lenticels. The latter serves as passages for gaseous exchange. Oxygen and carbon dioxide are channeled through these structures. The superficial roots are thus sensitive. The trees can easily die if the ventilating organs (lenticels) are blocked. Mangroves grow best where there is regular tidal flushing. Changes in hydrology resulting from dredging or blockage of natural tidal channels can severely reduce their growth and may result in death (**Figure 2**).

Some parts of the barrier forests, common along the Nigerian shoreline, occur within the field. They appear as small forest islands within the "sea" of mangroves. Typical plants within these forests include the breadfruit tree (*Artocarpus communis*), Abura (*Hallea ciliata*), *Allanblackia floribunda,* Fig tree (*Ficus trichopoda*), *Uapaca heudelottii*, and *Symphonia globulifera.* These species are typical of freshwater swamp forest. These forest islands within the sea of mangrove forests are analogous to oasis within the desert. They are the only places the coastal communities get their potable water from. They are thus sensitive ecosystems with a unique function. Salt-water intrusion can easily kill the vegetation within these islands as the plants inhabiting them are not salt tolerant. The coastal communities also utilize the forest islands for agricultural purposes. Breadfruit and plantain are cultivated within these forests. The forests thus allow them to get staple food in an otherwise nonarable environment (**Table 4**).

Indices based on the proportional abundance of species provide a simplified approach to diversity measurement. These *heterogeneity indices* take into account both evenness and species richness to produce a single value, the Shannon Index of diversity (H^1). The diversity was analyzed using the Shannon diversity index (H^1), Margalef index of species richness (d), Equitability or Evenness index (J), and Dominance index (D).

S/No.	Scientific name	Common name	Growth form	Uses
1	*Acrostichum aureum*	Mangrove fern	Herb	Crayfish trap
2	*Alchornea cordifolia*	Christmas tree	Shrub	Medicinal
3	*Alstonia boonei*	Stool wood		Timber, medicinal
4	*Allanblackia floribunda*		Tree	Timber
5	*Ananas comosus*	Pineapple	Herb	Edible fruit
6	*Anthostema aubryanum*		Tree	Timber
7	*Artocarpus communis*	Breadfruit	Tree	Edible fruit
8	*Bulbophyllum oreonastes*		Epiphyte	Ornamental
9	*Carapa procera*		Tree	Timber
10	*Calamus decratus*	Rattan palm		Basket, mat, rope, fish trap
11	*Cathormion altissimum*		Tree	-
12	*Dalbergia ecastaphyllum*		Scrambler	Rope, erosion control
13	*Elaeis guineensis*	Oil palm	Tree	Wine, edible oil, thatch, kernel
14	*Ficus trichopoda*	Fig tree	Tree	
15	*Fimbristylis sp.*		Herb	-
16	*Fleurya ovalifolia*	Tropical stinging nestle	Herb	Weed
17	*Hallea ciliate*	Abura	Tree	Timber, joinery, flooring board, canoe, pestle, mortar, pestle, barrel
18	*Lycopodium cernum*		Epiphyte	-
19	*Mariscus ligularis*	Seaside sedge	Sedge	Erosion control
20	*Mucuna sloanei*	Horse eye bean	Climber	Soup condiment
21	*Musa sapientium*	Plantain		Fruit
22	*Pandanus togoensis*	Screwpine		Mat, basket
23	*Pentadesma butyracea*		Tree	Timber, canoe, margarine, comb
24	*Piptadeniastrum africanum*			Timber, canoes, railway sleeper, furniture
25	*Raphia hookeri*	Wine palm		Wine, gin
26	*Raphia vinifera*	Bamboo palm		Thatch
27	*Rhizophora mangle*	Dwarf red mangrove	Tree	Tannin
28	*Rhizophora racemosa*	Tall red mangrove	Tree	Tannin, timber, fuel wood, prop, railway sleepers
29	*Sacciolepis africana*		Herb	-
30	*Sphenoclea zeylanica*		Herb	-
31	*Spondias mombin*	Hog plum	Tree	Edible fruit,
32	*Symphonia globulifera*		Tree	Timber
33	*Syzygium guineensis*		Tree	Timber

S/No.	Scientific name	Common name	Growth form	Uses
34	*Uapaca heudelottii*		Tree	Timber, edible fruit
35	*Vitex grandifolioala*	Black plum	Tree	Edible fruit

Table 3.
Plant species list in Tunu field in south Forcados [31].

Figure 2.
Mangrove vegetation in Tunu field (Note: Nymphaea lotus *floating around the mangrove roots) [31].*

The Shannon diversity index is a measure of community stability or ecological robustness that makes any community able to withstand external perturbation with minimal disruption. Diversity values of <1.00 indicate heavily polluted, 1–3 indicate moderate pollution, while values > 3 indicate clean and unperturbed habitats. The equitability index measures how evenly distributed the individuals are among the species, while the Dominance index is the opposite of equitability, that is, when individuals are not evenly distributed among the species, but rather, only a few species have a concentrated number of individuals. The environment belongs to the category of moderate perturbation or pollution.

2.2 Mangrove fauna of the Niger Delta

The highest concentration of aquatic organisms in Africa is found in Nigeria's Niger Delta mangrove forest zones (**Figure 3**). The species include from plankton (phytoplankton and zooplankton), aquatic invertebrates (bivalves, crabs, mussels, periwinkles, hermit crabs, etc.), and land insects (beetles, butterflies, mosquitoes, ants, and termites) to vertebrates (monkeys, manatee, pygmy hippopotamus, python, etc.) (**Figures 4** and **5**) [32]. The enigma is that the region is yet to be formally recognized as a global biodiversity hotspot by international agencies like the International Union for the Conservation of Nature (IUCN). Biodiversity hotspots in Africa have commonly grouped together without recognizing the rich biodiversity across different locations. For example, the entire forest biodiversity in West Africa was grouped as "West African forests" [34], whereas individual countries in this region have a diverse and rich supply of biodiversity.

The Nigerian mangrove swamps harbor a great diversity of macro in- and epi-faun of which crabs and mollusks form the great majority in terms of biomass. The fauna

Scientific name	Common name	Abundance	Proportion (%)
Dalbergia ecastophylum	Corn vine	6	24
Chrysobalanus icoco	Coco plum	4	16
Paspalum sp.	Silt grass	2	8
Scleria verrucosa	Bush knife	1	4
Combretum racemosum	Christmas rose	3	12
Osbeckia tubulosa	Melastomatacea	1	4
Mariscus longibracteatus	Sedge	1	4
Acrostichum aureum	Aquatic fern	1	4
Scleria naumanniana	Bush knife	1	4
Lycopodium cernuum	Fern	1	4
Alchornea laxiflora	Christmas bush	1	4
Syzygtium guineense	Myrtaceae	3	12
Total abundance		25	100
Number of species		12	
Abundance		25	
Dominance index (D)		0.1296	
Shannon Wiener ($H^{1)}$		2.248	
Margalef index (d)		3.417	
Equitability (J)		0.9046	

Table 4.
Abundance and diversity of weed species commonly found around mangrove forests in the Niger Delta, Nigeria [5].

faces severe water and salt balance, siltation, desiccation, oxygen availability, weight, and temperature limitations, especially when attempting to invade the adjacent land. Presently, few quantitative estimates of Niger Delta mangrove swamp fauna exist, perhaps due to the difficulty of sampling among thickets of mangroves, deep mud banks, and semi-stagnant lagoons, in addition to the difficulty in obtaining reliable estimates of fauna, which spend much of their time in burrows or very active when on the ground. In many cases, the seasonal occurrence and distribution of most macrofauna in the swamp are related to the characteristic fluctuations in environmental parameters associated with the dry and wet seasons as well as stress from anthropogenic activities. Besides, the influence of substratum on faunal species distribution in the Nigerian mangrove swamp of Nigeria has been reported [35].

There are presently few quantitative estimates of Niger Delta mangrove swamp fauna in the literature. Studies of Buguma Creek, a mangrove creek in Rivers State, Nigeria, resulted in the documentation of Zooplankton made up of 37 taxa in the Phyla Arthropoda, which was the most dominant: Coelenterata (Cnidaria); Ctenophora; Annelida; Chaetognatha; and Chordata. Macrobenthic invertebrates comprised 68 taxa in the Phyla Arthropoda and Annelida (the most dominant), Nematoda, and Mollusca. Fish species made up of 20 taxa in the families Sciaenidae (the most dominant), Elopidae, Ariidae, Gobiidae, Serranidae, Carangidae, Lutjanidae, Haemulidae, Monodactylidae, Sphyraenidae, Polynemidae, Trichiuridae, Cynoglossidae, and

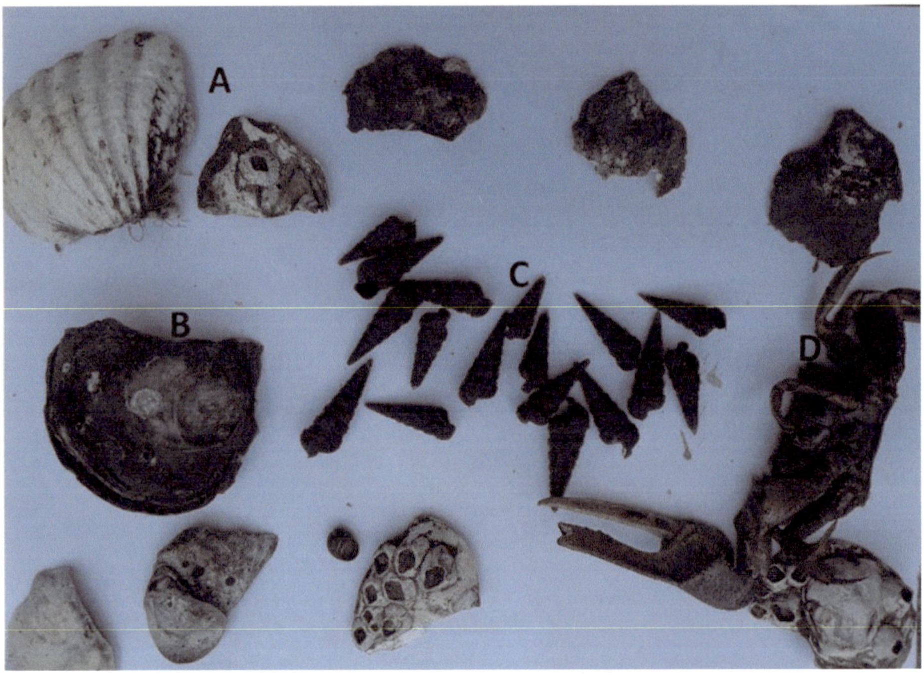

Figure 3.
Some species found within the Niger Delta mangrove forest: (A) Senilia senilis, *Anadara; (B)* Crassostrea gasar, *oyster; (C)* Tympanotonus fuscatus, *periwinkle; and (D)* Uca tangeri, *male crab [32].*

Figure 4.
Heslops pygmy hippo, Hexaproton *(syn-*Choeropsis*),* liberiensis heslopi.

Dasyatidae (these excluded members of the family Cichlidae (*Tilapia guineensis* and *Sarotherodon melanotheron*), Mugilidae (*Liza facipinnis, Mugilcephalus,* and *Mugil curema*), and *Sardinella maderensis* (Clupeidae), which the baited hooks and lines could not catch) [9].

Figure 5.
Protected python seen close to a residential building in Brass Island [33].

Table 5 shows the macrofaunal distribution and abundances in a study area that extends from the Cross River Estuary (which is relatively less perturbed and therefore used as a control swamp) to Takwa Bay in Lagos, approximately 870 km stretch and encompassing Imo River, Bonny River, Brass River, and Forcados and Escravos river estuaries. The preponderance of polychaetes in the upper ranks can be seen with *Capitella capitata, Chaetozone setosa,* and *Magelona filiformis* dominating. *Echinocardium* was ranked 65th with a density of 5/m^2. The coefficient of variation for each species is also presented in the table. This index (standard deviation ÷ mean density) indicates the spatial evenness in the abundance of the species, with numbers much less than one indicative of a uniform distribution [35].

Rank	Specimen	No/m^2	Coefficient of variation
1	*Capitella capilata*	62	0.20
2	*Chaetozone setosa*	56	0.26
3	*Megalona filiformis*	45	0.32
4	*Exogone lebes*	41	0.46
5	*Nemetina*	38	0.54
6	*Abra alba*	25	0.42
7	*Uca tangeri*	21	0.85
8	*Ampellisca*	18	0.83
9	*Nematoda*	16	0.67
10	*Nephthys* sp.	15	0.64
11	*Glycera*	9	0.38

Rank	Specimen	No/m^2	Coefficient of variation
12	*Scoloplos armiger*	8	0.62
13	*Polydora* sp.	7	0.56
14	*Turritella communis*	6	0.63
15	*Notomastus filiformis*	5	1.08
16	*Cirratulus cirratulus*	5	1.1
17	*Neries* sp.	5	0.68
18	*Marphysa belli*	4	1.32
19	*Hydrobia*	4	0.52
20	*Tellina* sp.	3	0.86
30	*Pectinaria*	5	0.65
35	*Pchygrapsus gracillis*	5	0.72
40	*Sesamaelegans*	15	0.68
48	*Sesarina alberti*	12	0.65
55	*Metagrapsus curratus*	12	0.65
60	*Littonia* sp.	2	1.21
62	*Mya arenaria*	1	1.02
65	*Echinocardium* sp.	1	1.23

Table 5.
Rank and abundance of dominant macrobenthic Fauna in the Niger Delta mangrove ecosystem during 1985–2005 [35].

3. Ecosystem services of Niger Delta mangroves

The four categories of ecosystem services that have been identified are: regulating services, provisioning services, cultural services, and supporting services.

Figure 6.
Prop roots providing attachment sites for oyster Crassostrea gazar *[18].*

The Niger Delta mangroves perform almost all the services listed above, such as from atmospheric and climate regulation, flood and erosion control, wood and timber for cooking fuel and construction to benefits such as esthetic value, sacred sites, traditional medicine, and supporting services such as nutrient cycling and habitat for fish nursery [18].

In addition to the aforementioned ecosystem services, the Niger Delta mangrove provides niche types that support the existence of a diverse fauna of all developmental stages. Several burrows contain diverse species of polychaetes, crabs, shrimps, and fishes supported by the rich food sources peculiar to the mangrove. Stilt roots of Mangrove plants provide substrate for the attachment of edible epifaunal organisms such as the oysters *Crassostrea* spp. (**Figure 6**) [18].

The mangrove trees conserve water resources and serve as windbreaks in many communities. Specifically, in the Niger Delta, other uses of mangroves by the indigenous people include building materials, food baskets, fishing tools, and so on [5].

4. Threats to mangroves of the Niger Delta

The mangrove forests of the Niger Delta are undergoing gradual but steady loss, resulting from uncontrolled deforestation for sand dredging and canalization. The mangrove forest is also cut to recover stems, which are used in producing firewood and wood for the construction of houses. Numerous oil and gas exploratory activities all over the Niger Delta area also open up the forests to further exploitation of resources [34]. Invasion by foreign species, such as nipa palms (*Nypa fruticans*), also threatens the existence of the rich indigenous mangrove species. Habitat loss is one of the three factors responsible for the recent extinction of species. Overexploitation and the introduction of exotic species are the other two. The consequence of the loss of mangrove habitat is the loss of ecosystem services it renders to society [34].

Oil exploration and extraction activities constitute Nigeria's most severe threat to the mangrove forest ecosystem. Nigeria's Oil Industry is located mostly in the mangrove forest ecosystem. The activities of various oil exploration companies have resulted in fragmentation, deforestation, and degradation of the mangrove forest ecosystem. The Nigerian Oil Industry has deforested mangrove ecological zones for drilling purposes and further deteriorated the health of the surrounding mangrove areas through oil spillages. Oil spill kills plants and animals in the estuarine zone, endangers fish hatcheries in coastal waters, and contaminates the flesh of commercially valuable fish. The oil that settles on beaches decimates the inhabiting organisms, while that settling on the ocean floor smothers and kills benthic (bottom-dwelling) organisms like crabs and disrupt major food chains. Oil also submerges birds, impairing their flight or reducing the insulating property of their feathers [36].

5. Management and conservation of the Niger Delta mangrove ecosystem

5.1 Mangrove forests

The Niger Delta mangrove forest requires urgent protection due to the adverse impact of anthropogenic activities resulting in the decimation of their populations in many regions. Incessant destruction of the mangrove forest will result in the local extinction of these species. Therefore, to reduce the anthropogenic impact on mangroves, there is an urgent need to establish protection and restrict human entry into the forest. Based on the importance of the biodiversity hotspots to the environment, it is pertinent to protect them for future generations [34].

In the past, the conservation of biodiversity hotspots was mainly focused on species richness, but now it considers ecosystem integrity, water quality, climate impacts, unique adaptations, ecosystem services, intact fauna and flora, specialized/unique habitat, and ecological processes.

Mangrove forest protection should follow the principles of reserve design, which include: the protection of entire habitats (because the more protected habitat, the better it is); avoidance of reserve fragmentation by anthropogenic activities such as the construction of highways through mangrove forests; establishment of the clumped reserve, which is far better than linear, for easy migration of species within the mangrove forests; and establishment of a circular reserve to minimize edge effects.

A damaged mangrove ecosystem could also be repaired through ecological restoration, the process of repairing damage caused by humans to the diversity, and dynamics of the indigenous ecosystem. This process includes erosion control, reforestation, removal of non-native species such as *N. fruticans* and weeds, revegetation of disturbed areas, daylighting streams, reintroduction of native species, and range improvement for targeted species. Hydrological links to natural restoration sites are also significant in allowing saline water inflow and mangrove seeds. The inflow of water also cleans the site from oil spillage and other pollutants [37].

There is an urgent need for a long-term management plan for the sustainable use of mangrove ecosystems. Monitoring, research, and evaluation constitute vital components of a successful conservation strategy for natural, managed, and man-made mangrove forests.

Conservation of existing mangrove forests is often more effective than planting new forests. When a decision for planting has been made, there is a need to emphasize the careful selection of appropriate sites and species and an ecosystem-based approach to mangrove planting and management, which utilizes and supports natural regeneration and other natural processes. In the process of any rehabilitation intervention work, planting should be done only when absolutely necessary. Local communities must be engaged in mangrove management to effectively maintain and enhance the protective function of the mangrove forest while providing a livelihood for local people and contributing to better assessment and governance of natural resources. Continuous assessment of mangrove forest status through research, economic assessment, and valuation is important for better conservation, planning, and management [38].

As a result of the current threat to forest resources, forest reserves have been created by the Nigerian government in the swamp/mangrove ecosystems of some Niger Delta States (**Table 6**).

5.2 Game reserves/wildlife sanctuary

Most of the Protected Areas are Forest Reserves. A few have been designated at one time or the other Game reserves (**Table 7**). None has the legal instrument for the designation as Game Reserve. They were recognized as having rich faunal compliments but never transited legally into Game Reserves.

A Preliminary Assessment of the Context for Reducing Emissions from Deforestation and Forest Degradation (REDD) in Nigeria commissioned by the Federal Ministry of Environment, the Cross River State's Forestry Commission, and UNDP Strict Nature Reserve (SNR) remarked that:

State	Swamp/Mangrove (ha)	High forest (ha)	Derived savannah (ha)	Total (ha)
Abia	2,870.00	4,949.60	1,302.50	9,122.10
Akwa Ibom	31,080.00	777.00	-	31,857.00
Delta	29,345.42	6,608.00	-	35,953.42
Edo	-	565,035.00	-	565,035.00
Imo	-	1,345.30	211.81	1,557.11
Rivers	-	121,440.00	-	121,440.00
Total forest type	63,295.42	700,154.90	1,514.31	764,964.63
%	8.274294	91.52774815	0.197958172	

Source: Federal Department of Forestry, 1998.

Table 6.
Forest reserves according to vegetation type.

S/N	Name	Habitat	Area (ha)	Remarks
Abia	NA	None	-	No designated game reserves
Akwa Ibom	Stubbs Creeks	Swamp forest	21,000	Stubbs creek is not managed as a game reserve
Bayelsa	Taylor Creek	Swamp forest	30,000	PA shared between Bayelsa & Rivers States. The larger part of Bayelsa state
Delta	Kwale	Degraded	340	Cassava farming and cattle trespass
Edo	Gele-Gele	Degraded	36,200	Retained as Forest Reserve. Lack of political will to transform FR to Game Reserve.
	Ohosu	Degraded	47,100	Never transformed to Game Reserve
	Okomu	Moist forest	11,200	Most of the original Okomu forest reserve has been sold to tree crop plantation developers. The core of the reserve has been preserved as a National Park and ceded to the Federal Government of Nigeria
	Ologbo	Degraded	19,440	Converted to an oil palm plantation
	Orle River	Converted	5,440	De-reserved
Imo	NA	-	-	No game reserve
Rivers	Taylor Greek	Swamp forest	30,000	Split between Bayelsa and Rivers states. No management regime in place

Source: Federal Department of Forestry, 1998.

Table 7.
Existing and proposed game reserves/wildlife sanctuaries in the zone.

These former forest reserves have been specifically designated for conservation, management, and propagation of wild animals in addition to the protection and management of the critical habitats on which they depend. In practice, almost all the game reserves in the country (except for Yankari, which was formerly a National Park) are highly degraded due to poor management; they best be described as "paper" game reserves.

6. Surface water and sediment quality in mangrove ecosystem

The Escravos Beach within the Niger Delta Mangrove was used as a case study [39]. The mean values of all physicochemical parameters investigated during the wet and dry seasons are summarized in **Table 8**. The water bodies sampled in the location had slightly alkaline pH in the wet season and slightly acidic in the dry season, high conductivity and dissolved solids, slightly high turbidity and color, and moderate dissolved oxygen and Biochemical Oxygen Demand (BOD) level. The water was brackish with higher salinity values in the dry than in the wet season. The cations were dominated by sodium and potassium, followed by calcium, truly reflecting the brackish nature of the water body. The nutrient level was higher in the wet than the dry season, probably due to the influx of allochthonous organic materials and ions into the water body. The parameters were all within the normal range stipulated by the Department of Petroleum Resources (DPR) (now called Upstream Petroleum Regulatory Commission, NUPRC) and Federal Ministry of Environment (FMEnv), except in a few cases such as turbidity, color, and total dissolved solids (TDS), where the limits were exceeded. The turbidity and color were expected to be higher in the wet season due to the input of organic materials from the terrestrial environment. The high total dissolved solids followed the electrical conductivity pattern, which was greatly influenced by the high salt content. The hydrocarbon pollution indicators, like the total petroleum hydrocarbons (TPH), polyaromatic hydrocarbons (PAH), BTEX (benzene, toluene, ethylbenzene, and xylene), and phenols, were all very low in concentrations and fell within the recommended limits by DPR and FMEnv. The low concentrations of heavy metals indicate no serious pollution from petroleum and other anthropogenic activities at the study location.

Microorganisms are responsible for forming various sediment and mineral deposits and dominating secondary waste treatment. Microorganisms such as bacteria and fungi act as living catalysts enabling numerous chemical processes in water and soil. Most significant chemical reactions in water, especially those involving organic matter and oxidation-reduction processes, occur through bacterial intermediaries.

The bacterial and fungal counts of the surface water samples from the study area were low. The total heterotrophic bacteria counts ranged between 0.7×10^3 cfu/ml and 2.4×10^3 cfu/ml. The total heterotrophic fungal counts ranged from nil to 1.0×10^3 cfu/ml. The hydrocarbon-utilizing bacterial and fungal counts were absent for most surface water samples.

The summary of sediment quality parameters is presented in **Table 9**. The sediment was moderately acidic (<7.0). The nutrient concentrations were higher in the sediment than in the surface water. The earth metals (cations) were low in concentration, with calcium dominating in the wet season. The heavy metals were low in concentration in all the stations. The higher sediment concentration is expected since the bottom sediment typically serves as a reservoir for heavy metals in the water. The recorded values are all below the intervention values. Concentrations above the intervention values corresponded to severe contamination. The target values indicated the soil quality levels ultimately aimed for [39].

There was no adverse effect of anthropogenic activities on the water and sediment of the study area.

The microbial load in the sediment was low for both the heterotrophic and the hydrocarbon-utilizing bacteria and fungi, indicating no serious hydrocarbon contamination.

Parameter	Unit	Dry season			Wet season			DPR, FMEnv limits
		Min	Max	Mean	Min	Max	Mean	
Physical								
Temp	0C	29.60	34.50	32.871	21.2	29.6	27.629	25,35
Color	Pt. Co	15.30	20.90	17.757	25.1	50	36.914	7
Turbidity	NTU	7.20	11.20	8.900	10.3	36	20.300	10,-
TSS	mg/L	8.00	15.00	11.000	10	36	19.286	30,30
TDS		11872.0	14687.0	13233.6	171	3780	1564.00	2000,-
pH		7.31	7.80	7.653	5.6	7.4	6.643	6.5–8.5,
EC	µS/cm	22400.0	27710.0	25174.3	362	7580	3135.3	
Salinity	‰	10.13	12.53	11.386	0.2	3.4	1.429	
DO		4.3	4.90	4.586	3.6	4.7	4.171	-,5
BOD5		2.8	4.40	3.471	1.7	3.9	2.500	10,10
COD		61.6	75.20	67.543	55.2	528	134.743	10,40
HCO3		171	281.00	222.429	73.2	201.3	108.057	
Na		334.4	467.90	438.043	10.9	221.6	100.457	20,-
K		371.6	551.10	491.529	3.3	98.1	60.286	
Ca		122	188.00	153.143	58	946	523.429	
Mg		87	134.00	117.143	23	722	319.571	75,-
Hardness		213	320.00	270.286	81	2960	1285.429	
Cl-		6990	8320.00	7587.143	63	198	129.429	600,200
PO4		0.46	2.37	1.520	5.3	9.2	6.800	5,-
NH4N		2.4	3.69	3.074	6.1	7.4	6.771	0.2,0.2
NO2-N		1.07	3.60	2.159	1.2	10.4	5.357	
NO3-N		2.17	4.17	3.039	2.5	19	9.929	20,-
SO4		336.1	377.2	361.0	50.5	253	123.7	200,-
H2S		0	0.00		0	0		1,20
THC		0.22	0.49	0.300	0.06	0.36	0.189	
TOC	%	0.59	1.20	0.804	0.39	1.19	0.860	
TPH	mg/l	0.079	0.15	0.112	0.059	0.137	0.094	
PAH	µg/l	0.02	0.10	0.047	0.017	0.089	0.037	
Benzene		0.002	0.06	0.013	0	0.013	0.002	
Toluene		0.001	0.00	0.002	0	0	0.000	
Ethylbenzene		0.001	0.01	0.003	0	0.002	0.001	
Xylene		0.001	0.00	0.002	0.001	0.003	0.002	
Phenol		0.004	0.02	0.009	0.002	0.012	0.006	
Heavy Metals								
Fe	mg/L	0.061	0.25	0.123	0.02	0.17	0.066	
Cu		0.051	0.16	0.112	0.04	0.14	0.107	

Parameter	Unit	Dry season			Wet season			DPR, FMEnv limits
		Min	Max	Mean	Min	Max	Mean	
Mn		0.016	0.10	0.059	0.01	0.09	0.056	
Zn		0.012	0.19	0.066	0.01	0.05	0.026	
Cd		0.002	0.00	0.003	0.01	0.05	0.021	
Cr		0.044	91.00	15.225	0.04	0.09	0.073	
Pb		0.026	0.05	0.039	0.02	0.08	0.049	
Ni		0.03	0.05	0.044	0.01	0.06	0.026	
V		0.011	0.04	0.027	0.01	0.02	0.012	
Hg		0	0.00		0	0		
Ba		0	0.00		0	0		
As		0	0.00		0	0		
CN		0	0.00		0	0		

Table 8.
Summary of physicochemical characteristics of surface waters in the Escravos Beach study area [39].

Parameter	Unit	Dry season			Wet season		
		Min	Max	Mean	Min	Max	Mean
pH		3.43	5.37	4.57	3.3	6.9	5.029
EC	µS/cm	2190	7650	3952.86	1070	1960	1660.00
Organic carbon	%	0.47	2.57	1.81	1.4	3.05	1.921
Total nitrogen	%	0.22	1.23	0.86	0.12	0.28	0.174
TOC	%	1.37	7.63	5.36	4.2	9.06	5.714
Na	meq/100 g	5.84	8.47	7.23	0.86	1.26	1.023
K		3.98	5.01	4.50	0.13	0.3	0.191
Ca		2.11	3.21	2.59	1.6	4.64	2.377
Mg		2.44	3.95	3.14	1.1	2.72	1.556
CEC		14.85	19.3	17.32	3.69	8.76	5.147
EA		1.6	3.1	2.31	0.9	1.6	1.214
CL-	mg/kg	48.5	132	82.39	13.6	20.2	17.357
PO4		12.2	17.5	14.20	16.2	23.7	19.557
NH4N		15.6	28.5	22.80	11	21.1	16.243
NO2-N		7.8	12.4	9.87	7.9	14.3	10.500
NO3-N		9.9	14.2	11.93	11.6	20.7	16.043
SO4		45.4	92.1	60.67	15.2	27.6	22.443
Silt	%	3.3	6.4	4.07	2.5	3.2	2.971
Clay		3.3	6.3	5.37	6.8	9.4	8.043
Sand		90.2	91.2	90.56	87.6	90.2	88.986

Parameter	Unit	Dry season			Wet season		
		Min	Max	Mean	Min	Max	Mean
THC	mg/kg	15.8	66.9	45.14	23	48.8	39.200
TPH		0.1	0.357	0.19	0.1	0.357	0.214
PAH		0.015	0.079	0.04	0.179	0.343	0.265
Benzene	µg/kg	0	0.007	0.00	0	0.007	0.003
Toluene		0	0	0.00	0	0.001	0.000
Ethylbenzene		0	0	0.00	0	0	0.000
Xylene		0	0.007	0.00	0.002	0.008	0.005
Phenol		<0.001	<0.001				
Heavy Metals							
Fe	mg/kg	40	74	53.71	230	554	381.143
Cu		0.6	22.4	21.00	14.6	15.7	15.100
Mn		0.16	0.24	0.20	28.6	34.6	31.214
Zn		3.6	23	14.73	40.1	55.9	48.829
Cd		0.04	0.07	0.06	1.2	2.2	1.757
Cr		0.6	2.06	0.97	1.2	3.6	1.957
Pb		1.86	3.82	3.32	1.1	5.9	2.829
Ni		0.21	0.75	0.39	1	2.5	1.657
V		0.144	0.51	0.27	0.6	1.8	1.143
Hg		ND	ND		0	0	
Ba		ND	ND		0.003	0.015	0.009
As		ND	ND		0.001	0.004	0.002

Table 9.
Comparison of the characteristics of sediment samples from the study area during the dry and wet seasons [39].

The mean values of physical and chemical parameters for surface water and sediment are true reflections of the typical mangrove quality unless there is gross contamination from oil exploration and production activities by oil companies.

7. Conclusion

The coastal areas of West Africa contain some of the world's richest ecosystems, including extensive mangrove forests, lagoons, and estuaries that support wide biodiversity with significant economic benefits to the coastal populations. Wetlands are important buffers for adjacent marine ecosystems, trapping sediments, nutrients, and many anthropogenic chemical contaminants. Mangrove ecosystems, for instance, serve as sinks for heavy metals because mangrove sediments' physical and chemical properties allow them to sequester large quantities of metals. There is wide variability in the ability of mangrove plants to absorb heavy metals because the sediment conditions vary widely. Different species of mangrove trees have different sediment-trapping root systems adapted to combating the anoxic conditions in the peat, including the pores or lenticels on the prop roots of *Rhizophora* spp. or on the pneumatophores (peg roots) of *Avecinnia germinans.*

Ineffective planning and management of coastal zone, the absence of environmental impact assessment, inadequate incorporation of environmental issues into feasibility projects, population pressures, as well as economic expansion are threatening the integrity of natural ecosystems. The main types of human activities that damage coastal ecosystems are (1) overexploitation, (2) physical alterations and habitat loss, (3) pollution, (4) introduction of alien species, and (5) global climatic change. In many coastal areas of Africa, vast areas of coastal wetlands and beaches have been cleared to make way for coastal development, industrialization, aquaculture, and agriculture. It destroys rich mangrove and estuarine fisheries and many commercially valuable species that depend on these coastal habitats as nursery grounds. Continuous examination of the relationship between man, the coastal environment, and its resources (human impacts) is clearly needed to avoid the risk of depletion and damage.

The rich ecosystem services of the Niger Delta mangrove, which is currently threatened by anthropogenic influence, require urgent protection and conservation attention. There is an urgent need for policymakers, development planners, and other stakeholders to arouse public consciousness for the restoration, rehabilitation, and conservation of mangrove ecosystems in Nigeria to achieve sustainable utilization.

Acknowledgements

This contribution contains information gathered from our Ph.D. research and environmental consultancy reports carried out for some oil and gas industries operating in the Niger Delta region of Nigeria. This is in addition to the cited literature. The authors wish to particularly acknowledge MacGill Engineering and Technical Services Limited, an environmental consultancy outfit that funded the Environmental Impact Assessment studies for South Forcados Development Project and the Escravos Beach Project carried out for SPDC.

Author details

Anthony E. Ogbeibu[1*] and Blessing J. Oribhabor[2]

1 Faculty of Life Sciences, Department of Animal and Environmental Biology, University of Benin, Benin City, Nigeria

2 Department of Fisheries and Aquatic Environmental Management, University of Uyo, Uyo, Akwa Ibom State, Nigeria

*Address all correspondence to: ogbeibu.anthony@uniben.edu

References

[1] World Bank. Overview. Available from: www.worldbank.org/en/country/nigeria/overview#1 [Accessed: October 13, 2019]

[2] FDF and FMEnv. National Forest Reference Emission Level for the Federal Republic of Nigeria. 2019. Available from: https://redd.unfccc.int/files/2019_submission_frel_nigeria.pdf

[3] Davies RM, Davies OA, Abowei JFN. The status of fish storage technologies in Niger Delta Nigeria. American Journal of Scientific Research. 2009;**21**:55-63

[4] Olalekan A, Gordon M. The Niger Delta wetlands: Threats to ecosystem services, their importance to dependent communities and possible management measures. International Journal of Biodiversity Science, Ecosystem Services and Management. 2011;**1**:1-19

[5] Numbere AO. Mangrove species distribution and composition, adaptive strategies and ecosystem services in the Niger River Delta, Nigeria. In: Sharma S, editor. Mangrove Ecosystem and Ecology Function. London, UK: IntechOpen; 2018. pp. 1-24. DOI: 10.5772/intechOpen.79028

[6] World Bank. Defining an Environmental Strategy for the Niger Delta. Nigeria: World Bank Industry and Energy Operations Division, West Central Africa Department; 1995. p. 76

[7] Okonkwo CNP, Kumar L, Taylor S. The Niger Delta wetland ecosystem: What threatens it and why should we protect it? African Journal of Environmental Science and Technology. 2015;**95**:451-463

[8] Ayanlade S, Drake N. Forest loss in different ecological zones of the Niger Delta, Nigeria: Evidence from remote sensing. GeoJournal. 2015;**81**(5):717-735. DOI: 10.1007/s/10708-015-9658-y

[9] Oribhabor BJ, Ogbeibu AE. Brackish Water Ecology of a Tidal Creek: Physical, Chemical and Biological Characteristics of a Tidal Creek. Germany: Lap Lambert Academic Publishing; 2012. p. 273. DOI: 978-3-659-28820-3

[10] Sheridan P, Hays C. Are mangroves nursery habitat for transient fishes and decapods? Wetlands. 2003;**23**(2):449-458

[11] Burke L, Kura Y, Kassem K, Revenga C, Spalding M, McAtlister D. Pilot Analysis of Global Cosystems: Coastal Ecosystems. Washington, D.C. USA: World Resources Institute; 2001

[12] IUCN-UNEP-WWF. World Conseration Strategy. Living Resources for Sustainable Development. IUCN–UNEP–WWF; 1980. p. 72

[13] IUCN. Global Status of Mangrove Ecosystems. Commission on Ecology Paper No. 3, Reprinted from the Environmentalist 3, Supplement No. 3. Gland, Switzerland: International Union for the Conservation of Nature and Natural Resources; 1983. p. 88

[14] Niger Delta Biodiversity Project (NDBP). A United Nations Development Programme (UNDP) and Global Environment Facility (GEF) Funded Project. Nigeria: UNDP-GEF; 2012. p. 172

[15] USAID/Nigeria. Foreign Assistance Act 118/119 Tropical Forest and Biodiversity Analysis. USA: The Cadmus Group LLC and ICF; 2020. Available from: Publications/Categorvid/930.aspx

[16] Itehm U. Wildlife Conservation Society Cross River National Park (Oban Division) Annual Report: January–December 2018. Wildlife Conservation Society; 2019. Available from: https://nigeria.wcs.org/About-Us/

[17] Ajibola MO, Awodiran OO. Assessing wetland services in the Niger Delta, Nigeria. International Journal of Humanities and Social Science. 2015; **5**(1):268-277

[18] Akanni A, Onwuteaka J, Mulwa R, Uwagbae MA, Elegbede IO. The values of mangrove ecosystem services in the Niger Delta Region of Nigeria. In: Ndimele EE, editor. The Political Ecology of Oil and Gas Activities in the Nigerian Aquatic Ecosystem. Academic Press; 2018. pp. 387-437. DOI: 10.1016/B978-0-12-809399-3.00025-2

[19] Ibe AC. Coastline Erosion in Nigeria. Ibadan Nigeria: Ibadan University Press; 1998

[20] Egborge ABM. Biodiversity of Aquatic Fauna of Nigeria. Abuja: Natural Resources Conservation Council; 1993. p. 173

[21] Nigerian Institute for Oceanography and Marine Research (NIOMR). Diagnostic Survey of Small Scale Capture and Culture Fisheries in Lagos State, Nigeria. NIOMR Technical Paper no. 56. Nigeria: NIOMR; 1989

[22] Scott JS. Report on the Fisheries of the Niger Delta Special Area. Port Harcourt: Niger Delta Development Board; 1966. p. 109

[23] Nduaguba DC. Utilization of Remote Sensing Data for Mapping Aquatic Ecosystems of the Nigerian Coastal Area. A Report Prepared for the Artisanal and Inshore Fisheries Development Project. Rome: FAO; FI: DP/NIR/77/001, Field Document 1; 1963. p. 24

[24] IPIECA. Biological Impacts of Oil Pollution, Mangroves. IPIECA Report Series. Vol. 4. London: International Petroleum Industry Environmental Conservation Association; 1993. p. 20

[25] Odiete WO. Environmental Physiology of Animals and Pollution. Lagos: Diversified Resources Ltd; 1999. p. 205

[26] Faunce CH, Serafy JE. Mangroves as fish habitat: 50 years of field studies. Marine Ecology Progress Series. 2006; **318**:1-18

[27] Baran E, Hambrey J. Mangrove conservation and coastalmanagement in Southeast Asia: What impact on fishery resources? Marine Pollution Bulletin. 1998;**37**:431-440

[28] Blasco F, Saenger P, Janodet E. Mangroves as indicators of coastal change. Catena. 1996;**1996**(27):167-178

[29] Edokpayi CA, Lawal MO, Okwok NA, Ogunwenmo CA. Physico-chemical and macrobenthic faunal characteristics of Kuramo water, Lagos, Southern Nigeria. African Journal of Aquatic Science. 2004;**29**(2):235-241

[30] Asuk AS, Offiong EE, Ifebueme MN, Akpaso OE. Species composition and diversity of mangrove swamp forest in Southern Nigeria. International Journal of Avian & Wildlife Biology. 2018;**3**(2): 159-164. DOI: 10.15406/ijawb.2018.03.00078

[31] SPDC. Environmental Impact Assessment Study (EIA) of South Forcados Development Project. Report submitted by MacGill Engineering and Technical Services Ltd. Nigeria: SPDC; 1995

[32] Numbere AO. Mangrove habitat loss and the need for the establishment of conservation and protected areas in the Niger Delta, Nigeria. In: Carmelo Maria Musarella CM, Ortiz AC, Canas RQ, editors. Habitats of the World-Biodiversity and Threats. IntechOpen; 2019. DOI: 10.5772/intechopen.89623

[33] Ogbeibu AE. Biodiversity Status and Concerns (Fauna). Biodiversity Study of the Brass LNG Project Area. Nigeria: Brass LNG Limited; 2009

[34] Myers N, Mittermeier RA, Mittermeier CG, Da Fonseca GA, Kent J. Biodiversity hotspots for conservation priorities. Nature. 2000;**403**:853

[35] Ewa-Oboho I, Asuquo FE, Edet P, Emeh EJ, Oladimeji S. Mangrove ecosystem of the Niger Delta–distribution and dynamics. Journal of Environmental Systems. 2005-2006; **32**(2):145-172. DOI: 10.2190/ES.32.2.e

[36] Aju PC, Aju JA. Mangrove forests in Nigeria: Why their restoration, rehabilitation and conservation matters. African Journal of Environmental and Natural Science Research. 2021;**4**(1): 84-93

[37] Numbere AO. Mangrove restoration under different disturbances regime in the Niger Delta, Nigeria. In: Sharma S, editor. Mangrove Ecosystem Restora tion. London, UK: IntechOpen; 2021. DOI: 10.5772/intechopen.96127

[38] Schmitt K, Duke NC. Mangrove management, assessment, and monitoring. In: Pancel L, Köhl M, editors. Tropical Forestry Handbook. Berlin, Heidelberg: Springer; 2016. DOI: 10.1007/978-3-642-54601-3_126

[39] SPDC. Environmental Impact Assessment Study (EIA) of Escravos Beach Location Project. Report Submitted by MacGill Engineering and Technical Services Ltd. Nigeria: SPDC; 2013

Chapter 3

Perspective Chapter: Remarks on the Relationship of Mangrove Recruitment and Thrombolithic Development in Coastal Lagoons

David Alfaro Siqueiros Beltrones

Abstract

Thrombolithic formations are sedimentary structures generated by precipitation, consolidation, and lithification of sediments and caused by the evolution of cyanophyte mats that promote coastal ground formation. These, and unlithified thrombolithic platforms or pro-thrombolites, are closely associated with mangrove trees and forests that use them as an anchoring substrate. So, in contrast with the suggested role of mangroves as coastal ground formers, here, evidence is provided that supports the hypothesis that mangroves are actually opportunists that eventually colonize various phases of the evolution of thrombolites. Photographs of mangrove specimens found affixed on various pro-thrombolithic structures, including the first thrombolite platform described in 2005, constitute palpability for the above. Also, lithified structures have been identified as rhizoliths belonging to mangrove trees as evidence that colonization on thrombolite platforms has occurred also during earlier geological events. It is desirable that specialists on mangrove ecology consider analyzing and assessing this theory from their own perspective.

Keywords: colonization, coastal ground formation, cyanophyte mats, mangrove recruits, pro-thrombolites

1. Introduction

Because thrombolites and other microbialites are well represented in the ancient geological record, information regarding their presence and distribution is critical for the understanding of past environments, as well as current landscapes, and may also serve to detect paleoecological issues. Living thrombolites or pro-thrombolites developing today exerts important changes in the geomorphological development of coastal areas, including landscape changes and alteration of coastal water currents and upper tidal limits [1]. Moreover, fossil and living microbialites are important references for explaining midterm environmental changes, such as the processes leading to the formation of coastal lagoons, coastal ground formation, and establishment of mangrove forests [2].

2. Thrombolithic and pro-thrombolithic formations discovered in lacunar environments

Thrombolites are defined as biosedimentary structures generated by the entrapment, binding, and consolidation of sediments as a result of the metabolic activity and growth of microorganisms, particularly cyanobacteria, but unlike the structure of stromatolites, in thrombolites no lamination is present, instead, there is a conglomerated matrix of cemented sediments; hence, the term "trombo" meaning clotted [3–6]. As with stromatolites, thrombolites are not difficult to recognize, especially by the curious look of science students, who immediately observe certain irregularities in a landscape. However, the sedimentary structures described and discussed here-on required a bit more heuristic zest in order to be recognized, because of their peculiar forms of development.

In 2005, a serendipitous observation triggered an exploration that revealed that most of the beaches of Ensenada de La Paz (coastal lagoon), Baja California Sur (BCS) were fringed by extensive unlithified sedimentary platforms, or irregular solidified mud spills (**Figure 1**) that were deemed as "living (active) thrombolites" [7]. Thus, it was inferred that these sedimentary structures were generated also by the entrapment, binding, and consolidation of sediments as a result of the metabolic activity and growth of filamentous cyanophytes, and were later called pro-thrombolites, inasmuch they were considered precursors of thrombolithic formations.

The clotted sediments of the pro-thrombolithic platforms at La Paz lagoon exhibited inclusions of shell fragments, irregularly packed, and surrounded by a sheet of active cyanobacteria [1] that some thrombolithic platforms lack because of prolonged exposure or being covered by more recent sediments [8]. However, most

Figure 1.
First thrombolithic platform recorded at El Conchalito, La Paz lagoon in 2005 that shows seaward accretion.

pro-thrombolithic and many thrombolithic formations may still present a conspicuous sheet of filamentous cyanobacteria, mainly *Microcoleus chthonoplastes* Thuret ex Gomont and *Lyngbya aestuarii* Liebman ex Gomont. Other common cyanophyte taxa are *Spirulina* sp., *Oscillatoria* spp., *Calothrix*, and unicellular forms, such as *Chroococcus* sp. and *Aphanotece* sp. [1]. Also, a diverse arrangement of benthic diatoms living on the surface of the thrombolithic platforms was described [9]. However, further inspection of the clotted matrix showed no traces of embedded diatoms [2].

From a wider view, soft extensive benthic mats observed to be formed roughly by the same filamentous cyanobacteria species cover most of the intertidal and subtidal shores (**Figure 2a** and **b**) of the La Paz lagoon, BCS., Mexico [9]. These are associated with soft, conglomerated (vertically accreted) mats and more or less consolidated platforms.

Due to their rocky appearance with a dark greenish covering [1], these seaward extensions had been hitherto unnoticed, and may still be overlooked when covered by filamentous macroalgae (**Figure 2c** and **d**). The said sedimentary structures may be in the form of muddy extensions, lithified platforms (**Figure 3**), and/or fragments of assorted sizes resembling mudstone, but in all cases, they occur as seaward soft or hardground accretions. In the case of lithified structures, microscopic analysis of their clotted matrix indicated that they were thrombolithic in nature [1]. Although these were first described around the shores of La Paz lagoon; thrombolites (rocky) had been recorded earlier further north in BCS [10].

Figure 2.
a) Extensive conglomerated cyanophyte mats located at El Centenario in the La Paz lagoon, BCS. b) Cross-section of a cyanophyte mat from El Conchalito, La Paz lagoon. c) Panoramic view of an extensive uncovered pro-thrombolithic platform, showing an anchored individual of Laguncularia racemosa*. d) View of the same pro-thrombolithic platform covered by macroalgae.*

Figure 3.
Smooth lithified thrombolithic platform at El Mogote, La Paz lagoon.

Following the identification and description of these microbialithic structures [1, 9], several issues were addressed, such as How do they originate? Are they a product of lacunar processes? How do they evolve? Are they solely evidence of past microbial activity? Or do they play a current ecological/geological role? And mainly, the one concerning the importance of pro-thrombolithic processes in the formation of coastal lagoons [1]. All these and other issues were further supported with evidence from various localities. But the most recent concern addressed their role as the primary coastal ground forming agents, a function commonly attributed to mangroves, but which were actually used opportunistically by mangrove recruits as anchoring substrate [2, 8, 11].

2.1 Association of thrombolithic formations in mangroves ecosystems

The above observations comprise the inspection of several coastal lagoons in the NW region of Mexico (**Figure 4**). The shores of the explored coastal lagoons are characteristically populated by marsh vegetation, mainly *Salicornia* spp. and three species of mangrove: *Avicennia germinans* (Linnaeus) Stearn, *Rhizophora mangle* Linnaeus, and *Laguncularia racemosa* (Linnaeus) Gaertner, as mentioned above, closely associated to the pro-thrombolithic grounds. Particularly targeted were the ecosystems at La Paz lagoon and Bahía Magdalena, because in the former the initial observations were made and have been more closely studied. While in the latter it is where the (big-time) hypothesis that "thrombolites and pro-thrombolites should be found in other coastal lagoons" was first favorably contrasted. Eventually, the said structures were found later elsewhere, also associated with mangroves, including San Ignacio lagoon [1],

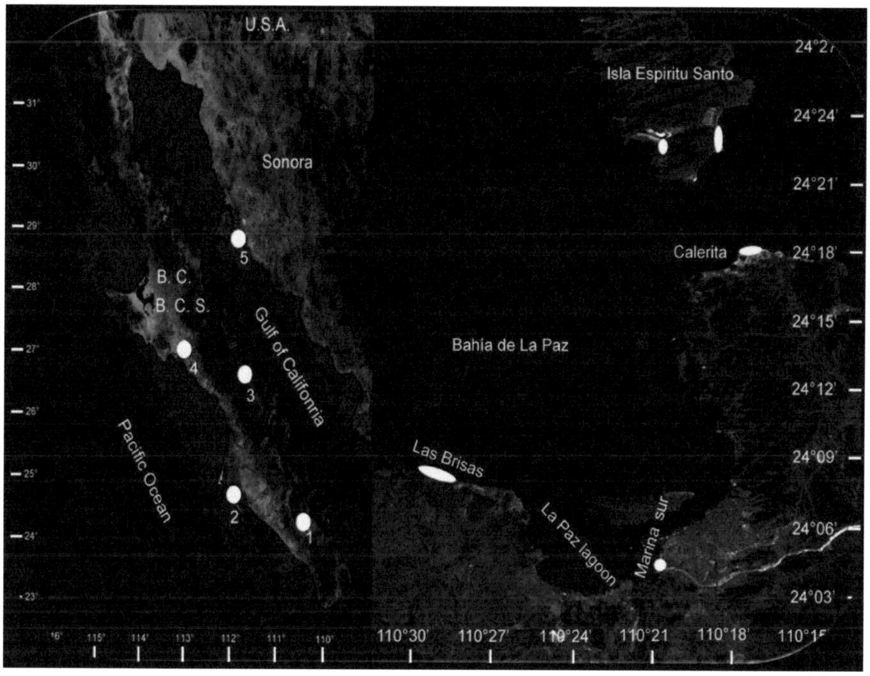

Figure 4.
Localities where observations on thrombolithic structures associated with mangrove forests have been made: 1) La Paz lagoon and surrounding sites (inset); 2) Magdalena bay; 3) Concepcion bay; 4) San Ignacio lagoon; and 5) Sonora.

Bahía Concepción (BCS), and more conspicuously in Sonora. Likewise, thombolithic platforms were found in extra-lacunar environments, where pre-ichnofossil evidence of past mangrove presence was also observed [2].

In the La Paz lagoon, located on southeastern (gulf) coast of the Baja California Peninsula, *Avicennia germinans* is the most common (black) mangrove though red mangrove (*R. mangle*) is also conspicuous, while white mangrove (*L. racemosa*) is scarce. In most cases, where thrombolithic and pro-thrombolithic platforms are present, mangroves are found associated with them mainly as anchoring substrate (**Figure 5a**). Also, consolidated pro-thrombolithic blocks of 50 cm high (approx.) were discovered (at Marina Sur) associated with well-established *A. germinans* trees, nearby the first platform observed in the lagoon (**Figure 5b**). These blocks were broken-off from a platform structure, and those more exposed to wave energy appeared whimsically eroded, while others seemed worn particularly around the lower part, reminding the bun shaped stromatolites of Shark Bay, Australia [2].

Also, because lagoon-like environments are evident on the island coves in the southern Gulf of California, explorations were done at Isla Espiritu Santo and Isla San Jose, where mangrove forests are well established [2]. Additionally, observations on thrombolithic-like platforms and blocks outside lacunar environments in the Gulf of California is graphically documented northwest (Las Brisas) of the La Paz lagoon [2], and to the north (Calerita), where two distinct geological levels of formation have occurred (**Figure 5c**) inside Bahia de La Paz in the southern Gulf. In this case, an issued null hypothesis stated that thrombolithic structures would not

Figure 5.
*a) First observed association between mangrove (*Avicennia germinans*) and a thrombolithic platform at El Conchalito, La Paz lagoon. b) Pro-thrombolithic platform, broken down into blocks, associated with mangrove trees at La Marina, La Paz, lagoon. c) Broken-off thrombolithic blocks from extensive platforms at Calerita, outside La Paz, lagoon in Bahia de La Paz, Mexico, which appears to have rested on an earlier pro-thrombolithic platform. d) Ichnofossils, allegedly of anchored mangrove roots on thrombolithic platforms at Calerita outside La Paz lagoon.*

be found, inasmuch as the explored sites were not lacunar environments. However, as mentioned above, although no mangroves currently occur in these two extra-lacunar localities, evidence of their past presence on the thrombolithic platforms as ichnofossils is present (**Figure 5d**).

In the case of Bahia Magdalena lagoon, located on the southwestern (Pacific) coast of the Baja California peninsula, its shores are densely populated by mangrove forests of the same three species, though much bigger, also closely associated with thrombolithic structures [2]. But where, unlike in the La Paz lagoon, *L. racemosa* is very common. There, several sites were explored, specifically around Puerto San Carlos where thrombolithic platforms were exposed (**Figure 6**), while structures assumed to represent transitional stages between conglomerated mats and thrombolites were recorded landward (**Figure 7a**) among and behind the mangrove trees line [2].

Observations on the coast of Sonora were based both on direct inspection and indirect evidence. In the first case, a quick exploration was carried out at Estero El Soldado in Guaymas, where the same conglomerated formations were noted along with thrombolithic-like platforms, also associated with mangroves [2]. However, the above initiative was triggered out of viewing a slide, during a symposium,

Figure 6.
First thrombolithic platforms recorded at San Carlos, Bahía Magdalena, BCS. These findings are backed up by the posed "big time" hypothesis on their expected occurrence within other coastal lagoons.

from Estero Santa Rosa, Sonora shows Seri natives seating on thrombolithic blocks within a copious mangrove environment (**Figure 7b**). The said blocks are quite similar to those found along the Malecon road at Bacalar, Quintana Roo, Mexico, where mangrove forests are extensive, even though more than 50% is reported to be have been destroyed. There, stromatolites have become celebrities since being described [12], but not thrombolites (**Figure 7c** and **d**) for which a single indirect study (on epilithic diatoms) is known [13]. Unfortunately, the Santa Rosa site has not yet been accessed and samples from either site in Sonora are not available, and neither from Bacalar where they lack the required minimum attention (**Figure 8**). Notwithstanding, in spite that sedimentological and other direct examinations are pending, the association of pro-thrombolithic formations and mangroves is clearly confirmed.

This constant correspondence between thrombolithic structures and mangroves suggested the hypothesized dependence of the latter. However, it was reinforced by the occurrence of actual recruiting of three mangrove trees (two *Avicennia germinans* and one *L. racemosa*) during the term of the present investigation on the first thrombolithic formation recorded (**Figure 9a**), where rhizolith-like structures were observed below the level of the current platform [11]. Likewise, another recruit of black mangrove was documented on the second recorded platform, where a much older specimen of *A. germinans* was present (**Figure 9b**), and by the incipient ichnofossils-like structures from El Mogote, the sand-bar defining the La Paz lagoon (**Figure 9c** and **d**). These complex structures seemed derived from the interaction between dead mangrove root systems and pro-thrombolithic

Figure 7.
a) Partially uncovered pro-thrombolithic platforms found behind the mangrove line at San Carlos, Bahia Magdalena. b) Thrombolithic blocks at Santa Rosa estuary, Sonora, associated with extensive mangrove forests. Image by Diana Luque (Ecoturismo Seri, CIAD and CtamCoyai, a.C.). c) Author and godson, with conspicuous fresh-water stromatolites peeking out from behind at Bacalar lagoon. d) the author on trombolithic blocks at Bacalar, Quintana Roo long ago excavated to make way for the local Malecon road.

activity and are most likely incipient rhizoliths. As mentioned above, in the northern part of BCS, in Bahia Concepción, fossil deposits include stromatolites and thrombolites [10] as well as rhodolith beds, muddy sandstone, rhizoliths, and mangroves [14].

2.2 Putative role of mangroves in coastal ground formation

The relationship between the role of mangroves and the morphodynamic response of the shoreline as it has been outlined [15], indicates that sedimentation modifies the geomorphological setup and influences the soil characteristics, groundwater reach, and substrate salinity, which determines mangrove zonation and species distribution. The said process includes deposition of fine-grained, clay-dominant particles within the forest floor, which is considered to be one of the driving factors of land-building and shoreline progradation. Likewise, coastal sedimentation is favored by mangrove vegetation resisting the tidal water flow and trapping the sediments through the network of their roots. In this way, mangroves, acting as traps for both mineral and organic sediments, control sedimentation and thus form their own survival ground, and eventually, the result permits a clear differentiation between coasts with and without mangroves [16].

Figure 8.
Faith for thrombolithic remains? And a consequence of lack of minimum care?

In accordance with the above, it has been commonly accepted that mangroves functioned as sediment traps and eventually causing ground formation [17, 18]. Although other observations have suggested that ground formation is actually a preexisting process that is accelerated by mangroves after colonizing suitable areas [15]. In general, however, the origin of ground available for mangrove recruiting is scarcely addressed. Here, mangroves are proposed as being opportunists that colonize an already available substratum, in this case, thrombolithic structures.

Our observations thus provide evidence that thrombolite and pro-thrombolite platforms are, along with thick cyanophyte mats, the substrates most likely to promote mangrove colonization [2], evolving from an organic mat, and passing through the stages of conglomerate mats, all constructed mainly by long multiseriate filaments of *Microcoleus chthonoplastes* having thick mucilaginous sheaths that remain long after death (**Figures 10** and **11a**), and pro-thrombolite platforms, into sedimentary rock by micrite deposition [1, 2].

During their evolution, the various thrombolithic phases are opportunistically colonized by mangrove recruits (**Figure 11b**, **c**, and **d**). It is thus a preexisting process that is accelerated by mangroves after colonizing suitable areas as suggested previ ously [15], such as those provided by the pro-thrombolite and thrombolite platforms.

3. Conclusion

The present thesis had a serendipitous origin. During an exploration for appropriate sites to collect epipelic diatoms in the La Paz lagoon, I came upon

Figure 9.
a) Updated images (2022) of successful mangrove recruiting of two Avicennia germinans *individuals and one* Laguncularia racemosa *on the first thrombolithic platform recorded at El Conchalito, La Paz lagoon, around 2015. b) Recruiting of an individual of* A. germinans *on the second thrombolithic platform recorded at El Conchalito, La Paz lagoon occurring around 2015. c) Conglomerate of ichnofossils-like structures of mangrove roots (rhizoliths) topped by a thrombolithic plate at El Mogote, La Paz lagoon. d) Incipient ichnofossils assumed to be rhizolith-precursors of mangrove roots from El Mogote, La Paz lagoon.*

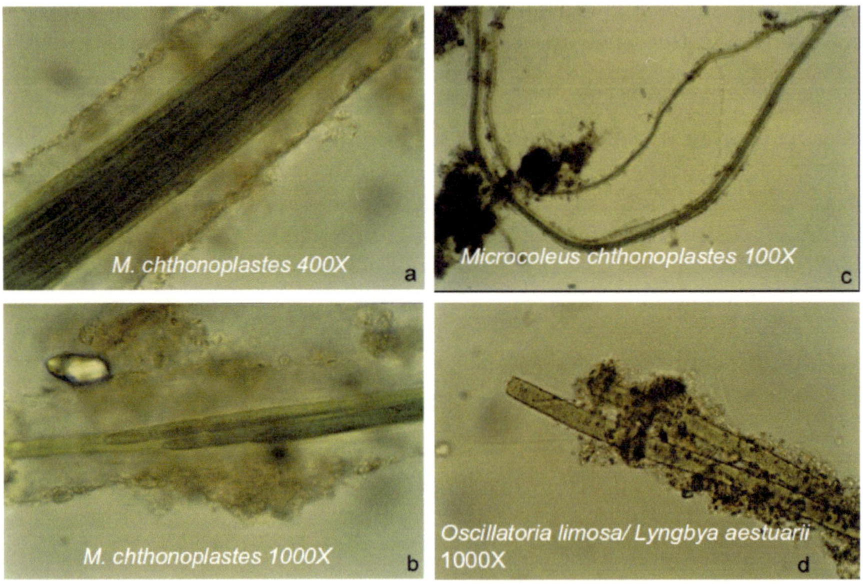

Figure 10.
a, b, and c) different magnifications for a multiseriate filament of Microcoleus chthonoplastes *showing thick mucilaginous sheath; and d)* Oscillatoria limosa. *Both were from a cyanophyte mat collected at El Conchalito, La Paz bay.*

Figure 11.
a) Close-up of empty Microcoleus chthonoplastes *sheaths from a desiccated mat behind the mangrove line at San Carlos, Bahia Magdalena. Microphotograph is taken at 100×. b) Recruit of* Rhizophopra mangle *affixed in a cyanophyte mat in front of the mangrove line in the intertidal of San Carlos, Bahia Magdalena. c) Recruit of* Laguncularia racemosa *anchored on a pro-thrombolithic (unlithified) block at El Mogote, La Paz lagoon. d) Recruits of* Rhizophora mangle *anchored on a seaward extended thrombolithic platform at San Carlos, Bahia Magdalena.*

an irregularity on the shore that looked like a disgusting land dump, darkened by extreme irradiance, which got me thinking about what was I to do in terms of the expected distribution of the diatom assemblages. Fortunately, a quick turn of thought focusing on the sedimentary structure gave a start to a series of questions and hypotheses based on the above observations that ended in the proposal of a complementary theory for the origin of coastal lagoons. Likewise, this was followed by yet another complementary theory, the present one on the origin of coastal ground that explained the availability of substratum for the opportunistic recruiting and establishment of mangrove forests. Albeit somewhat iconoclastic, it is still an incipient theory in need of much work to fill in the many gaps that individual creativity can observe. Let us hope that specialists on mangrove ecology and geomorphology of coastal lagoons consider this theory from their own perspective for proper continuance, conceiving ideas for giving way to various research problems and enriching it.

Acknowledgements

This study is the basis for a periodic presentation in class and seminars at Cicimar-IPN, addressing the role of scientific hypothesis in the scientification of serendipitous observations. During a great part of this investigation, several colleagues eagerly aided in gathering the evidence referred to above: Oscar U. Hernández Almeida, Uri Argumedo Hernández, Rubén García Gómez, and Janette Murillo Jiménez. I thank my colleague José Borges who recently rescued the draft of this paper from cybernetic mayhem. Francisco López Fuerte formatted the photographic material.

Author details

David Alfaro Siqueiros Beltrones
Department of Plankton and Marine Ecology, Interdisciplinary Center of Marine Sciences-Instituto Politecnico Nacional, Av. Instituto Politécnico Nacional S/N, México

*Address all correspondence to: dsiquei@ipn.mx

References

[1] Siqueiros Beltrones DA. Role of s-thrombolithic processes in the geomorphology of a coastal lagoon. Pacific Science. 2008;**62**(2):257-269. DOI: 10.2984/1534-6188(2008)62[257:ROPITG]2.0.CO;2

[2] Siqueiros Beltrones DA, Hernández Almeida OU, Murillo Jiménez J. Further remarks on the role of pro-thrombolites in the origin of coastal lagoons for northwestern México. Hidrobiológica. 2012;**22**(3):244-257

[3] Charpy L, Larkum AWD (Eds.). Marine cyanobacteria. Bulletin de l'Institut Océanographique, Musée Océanographique, Monaco. Numéro spécial. 1999;**19**:624

[4] Riding R. Microbial carbonates: The geological record of calcified bacterial algal mats and biofilms. Sedimentology. 2000;**47**:179-214. DOI: 10.1046/j.1365-3091.2000.00003.x

[5] Shapiro R. A comment on the systematic confusion of thrombolites. PALAIOS. 2000;**15**(2):166-169. DOI: 10.2307/3515503

[6] Stal LJ. Cyanobacterial mats and stromatolites. In: Whitton BA, Potts M, editors. The Ecology of Cyanobacteria; their Diversity in Time and Space. Netherlands: Kluwer Academic Publishers; 2000. pp. 61-120

[7] Siqueiros Beltrones DA, Argumedo Hernández U, Hernández Almeida OU. Trombolitos litificados dentro de la Ensenada de La Paz, B.C.S., México. CICIMAR Oceánides. 2006;**21**(1, 2):155-158. DOI: 10.37543/oceanides.v21i1-2.32

[8] Siqueiros Beltrones DA, Félix Pico EF, Hernández Almeida OU. Stratigraphic evidence of pro-thrombolithic ground formation around the La Paz lagoon (México). CICIMAR Oceánides. 2009;**24**(1):59-63. DOI: 10.37543/oceanides.v24i1.54

[9] Siqueiros Beltrones DA. Diatomeas bentónicas asociadas a trombolitos vivos registrados por primera vez en México. CICIMAR Oceánides. 2006;**21**(1, 2):113-143. DOI: 10.37543/oceanides.v21i1-2.30

[10] Miranda-Aviléz R, Beraldi-Campesi H, Puy-Alquiza MJ, Carreño AL. Estromatolitos, tufas y travertinos de la sección El Morro: depósitos relacionados con la primera incursión marina en la Cuenca de Santa Rosalía, Baja California Sur. Revista Mexicana de Ciencias Geológicas. 2005;**22**(2):148-158

[11] Siqueiros Beltrones DA, Murillo Jiménez JM, García Gómez RE. Observations supporting the hypothesis of the colonization of thrombolithic platforms by mangroves. Hidrobiológica. 2017;**27**(1):119-121. DOI: 10.24275/uam/izt/dcbs/hidro/2017v27n1/SiqueirosB

[12] Gischler E, Gibson MA, Oschmann W. Giant Holocene freshwater microbialites, Laguna Bacalar, Quintana Roo, Mexico. Sedimentology. 2008;**55**:1293-1309. DOI: 10.1111/j.1365-3091.2007.00946.x

[13] Siqueiros Beltrones DA, Argumedo Hernández U, Hernández Almeida OU. Diagnosis prospectiva sobre la diversidad de diatomeas epilíticas en Laguna Bacalar, Quintana Roo. México. Revista Mexicana de Biodiversidad. 2013;**84**(3):865-875. DOI: 10.7550/rmb.33960

[14] Johnson ME, Ledesma-Vázquez J, editors. Pliocene Carbonates and Related

Facies Flanking the Gulf of California, Baja California. México. USA: Geol. Soc. Am. Spec. Pap; 1997. p. 318

[15] Woodroffe C. Mangrove sediments and geomorphology. In: Robertson AI, Alongi DM, editors. Tropical Mangrove Ecosystem. Washington DC: American Geophysical Union; 1992. pp. 7-41

[16] Thampanya U, Vermaat JE, Sinsakul S, Panapitukkul N. Coastal erosion and mangrove progradation of southern Thailand. Estuarine, Coastal and Shelf Science. 2006;**68**(1-2):75-85. DOI: 10.1016/J.ECSS.2006.01.011

[17] Dawes CJ. Marine Botany. USA: John Wiley and Sons Ltd.; 1981. p. 628

[18] Kathiresam K. How do mangrove forests induce sedimentation? Revista de Biología Tropical. 2003;**51**(2):355-360

Chapter 4

Perspective Chapter: Mangrove Deforestation and Sustainability of Malacological Resource Exploitation in the Estuarine Ecosystem of the Saloum River Delta, Senegal

Hamet Diaw Diadhiou, Moustapha Deme and Djiby Thiam

Abstract

The estuarine and mangrove ecosystem of the Saloum River Delta is a fragile natural environment. It constitutes an essential element of the estuary's trophic enrichment characterized by the presence of multiple bolongs. The harvesting of marine mollusks (oyster, ark, cymbium) in the Saloum Delta is mainly done by adult women who represent more than 95% of the stakeholders in this sector. The human activities practiced around this ecosystem are multiple: agriculture, fishing, tourism, beekeeping, use of mangrove wood for cooking and housing, empty shells of arches in the construction of roads and houses, and the composition of aviculture food. The periods of drought in the past three decades, as well as the multiform aggressions of the environment and the populations on the environment, have profoundly modified the hydrological and bio-geochemical functioning of the mangrove ecosystem of Sine-Saloum and significantly degraded the stocks of marine mollusks.

Keywords: mangle, oyster, estuary, impacts, forest

1. Introduction

Among the objectives of the project "Valorization of Species for the Sustainable Use of Wildlife Resources in Senegal (VALUES)," implemented by IUCN in 1998, was the promotion of the sustainable use of flora, fauna, and continental waters by rural populations through appropriate national policies, planning, and investment. The case study on Mangrove deforestation and the sustainability of the exploitation of malacological resources in the estuarine ecosystem of the Saloum River falls within this framework. The mangrove ecosystem is a fragile natural environment.

It is an essential element in the trophic enrichment of the estuary. The roots of the mangroves trap the organic matter that they return to the sea. The decomposition of mangrove leaves also brings organic matter to the marine environment. The mangrove is also the biotope where most of the marine mollusk's harvested by local populations live. Mangroves play a primary role against coastal erosion by stabilizing the substrate and retaining sediments.

The human activities practiced around this ecosystem are multiple: agriculture, fishing, tourism, beekeeping, use of mangrove wood for cooking and housing, empty shells of arches in the construction of roads and houses, and composition of chicken feed (the calcium contained in the oyster shell powder makes the shell of the hen firmer). The chronic drought of the 1980s and its consequences on agriculture have created hyper-salinity and acidification of most soils in almost the entire watershed of the Sine-Saloum ecosystem and consequently the loss of biodiversity. At the socio-economic level, the disruptions caused by these series of degradation have led in recent years to a strong migration of people from inland to coastal areas and a drop in food production (agricultural, fisheries, and forestry) and income, especially for poor women, who are poorly equipped and less prepared for the intensification of exploitation and competition.

The mangrove case study provides a better understanding of the problem of the use of malacological resources in terms of the economic benefits that can be derived from them as well as in terms of the sustainability of use patterns by determining the economic and social values that the populations of the Sine-Saloum Delta attach to these resources. It attempts to assess the impacts of deforestation on the sustainability of the exploitation of marine mollusks in this environment. This study has specifically focused on: (i) proceeding to the description of the sector in order to characterize the actors, the levels of remuneration of the actors, the modes of valorization of the products, and the destination markets; (ii) identifying the level of mangrove deforestation in order to better assess the state of the resource and analyze the impacts in relation to the different uses; and (iii) assessing the impact of deforestation on the sustainability of marine shellfish harvesting and proposing management modalities.

2. Methodology

The study was based on existing data (literature review) and the collection of additional information from various mangrove stakeholders (women seafood gatherers, fish product processors, mangrove wood operators, tourism professionals, NGOs, researchers, etc.). The VALUES project database was also used. This approach made it possible to structure this study around the following axes: (i) assessment of the impacts of mangrove deforestation on the sustainable livelihoods of the various stakeholders; (ii) the typology of the different uses of malacological resources; (iii) estimation of the income that the populations derive from the exploitation of marine mollusks in the Saloum Delta; (iv) assessment of the impact of the disappearance of the mangrove forest on the living conditions of these actors in order to predict supporting actions of IUCN and other partners in a perspective of sustainability and viability of the exploitation of natural deposits of marine mollusks in the mangrove ecosystem of Sine-Saloum; and (v) the proposal of modalities for the management and sustainable use of the Sine-Saloum mangrove that could later be extended to other species and geographical areas.

3. Results and discussion

3.1 The Sine-Saloum mangrove ecosystem

The Sine-Saloum mangrove is located in the intertropical zone, between 12° 30′North and 16°30′North latitude and 11°30′West and 17°30′West longitude, 150 km south of Dakar (**Figure 1**). The Sine-Saloum estuary covers an area of about 80,000 ha (less than 0.6% of the mangrove area in this region). Its climate, of the Sudanese type, is characterized by two distinct seasons, a seven-month dry season (November to May) and a five-month rainy season (June to October). The average annual rainfall (1951–1980) in the basin was between 880 mm in the south and 480 mm in the north [1]. Across the basin, the overall average was 828 mm in the 1950s [2]. It fell to around 500 mm in the 1980–90s. The hydrographic system of this complex consists of three main rivers: the Saloum (110 km long) to the north and northeast, the Bandiala (18 km) to the south and southeast, and the Diomboss (30 km) in between. These rivers are bordered by bolongs. These rivers are bordered by intertidal mudflats more or less colonized by mangroves.

From Sangomar to Foundiougne, the Saloum channel is relatively deep (depths below 13 m are rare and, in some cases, exceed 25 m at the level of the trenches). Upstream from Foundiougne, the Saloum River crosses the Sine River and becomes very sinuous while keeping a general direction up to Kaolack. Its width is relatively small in this area, rarely exceeding 500 m. The depth also decreases, usually remaining below 5 m. The Diomboss channel is relatively deep. Funds of 10 m are regularly found there. Depths of up to 25 m are recorded in some trenches. Upstream, the Diomboss is divided into several tidal channels, locally called bolongs. Diomboss and Bandiala are characterized by a bolon network, extremely dense unlike the right side of the Saloum River. These bolongs, like the main arms, are bordered by intertidal mudflats. There is no permanent river in the Saloum Delta basin. The slope of the river is gentle, less than 0.6%. It is at the origin of the current result observed upstream. This phenomenon affects the sedimentology and hydrology of the environment [3]. This reverse mode of operation of the estuary governs the arrangement of the salt spindle and the trapping of fine particles.

A large part of the mangrove in this area, 60,000 ha, is located in the Saloum Delta Biosphere Reserve administered by the National Parks Department (DPN in French). The central part (the National Park) and part of the buffer zone are managed by the state, while the rest is managed by local communities. The park and biosphere reserve are located in the Sine and Saloum delta, and the area includes many sandy islets and lagoons, Sangomar Point with its coastal waters, and the Fathala forest. Most of the terrestrial area is covered with mangroves, forests, and saltwater streams. The mangrove that dominates here consists mainly of *Rhizophora racemosa, Rhizophora mangle, Rhizophora harissonni*, and *Avicennia nitida*.

The extension of rice cultivation and the exploitation of the forest in the Fathala region constitute a permanent threat to the mangrove [4]. The trophic richness of the Saloum estuary is related to the presence of the mangrove, the existence of very numerous bolongs, and the very flat relief due to the fact that vast areas are covered and discovered at each tidal cycle [5].

Due to the absence of floods in the Sine-Saloum River basin and a strong decrease in freshwater inflow, the trophic richness produced around mangrove vegetation hardly goes beyond its production area toward the adjacent coastal environment. As in Casamance, the main source of trophic enrichment in Sine-Saloum is the

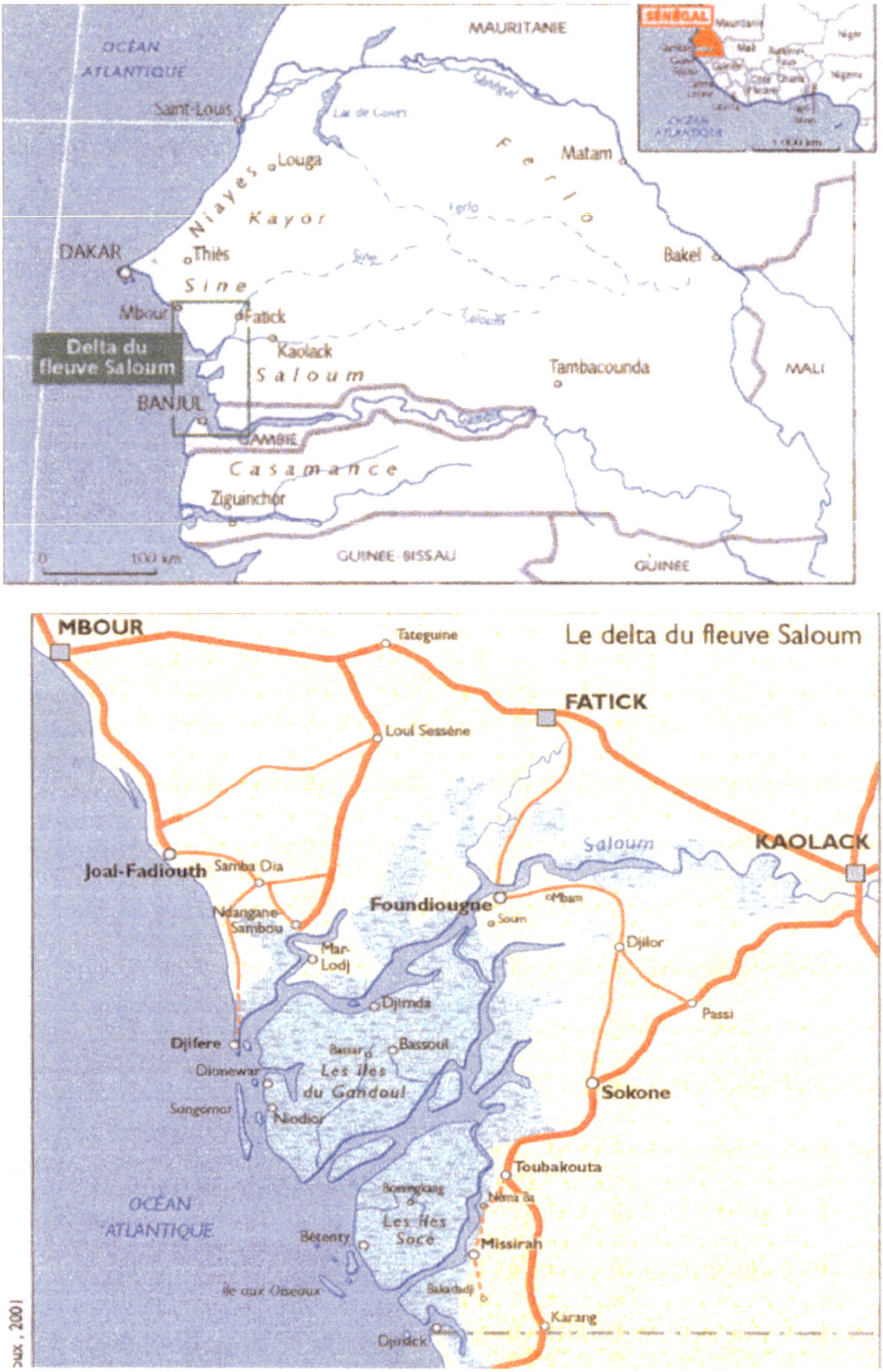

Figure 1.
Map 1: Location map of the Saloum delta, Senegal (Source: Master2 student dissertation).

remineralization of organic matter, which comes essentially from the mangroves [6].

The mangrove is very extensive in the southern part of Sine-Saloum, where it occupies a good place in the space between the bolongs (name designating small tributaries in the Mandinka language) [7]. Along the edges of the tidal channels located in the downstream part stands the high mangrove (mangroves, 7 to 11 m high). Bare or tanned salty soils are practically absent in this zone.

The shores of the Diomboss and the zone located east of Dionewar, in the central region of the Saloum delta, are occupied by tannes and a low mangrove (2 to 8 m high).

The north of the delta is characterized by a very degraded mangrove whose tree height does not exceed 4 m. This mangrove disappears upstream of the tributary of the Saloum.

The Mollusks of the Sine-Saloum estuary can be classified in the following particular biotopes in relation to the energy of the water movements:

- *Low-energy estuarine biotope* with the limit of the mangroves *Thais callifera*, *Thais forbesi*, and *Crassostrea gasar*. The main species exploited in this group is the mangrove oyster, *C. gasar*. The species necessarily fixes itself on a solid substrate. This type of substrate is first provided by the mangrove's aerial roots. However, oyster banks can be found on sandy or muddy sediments, provided that there are shells on the bottom allowing the fixation of larvae [8]. The species is present in the Saloum estuary, between Djifère and Foundiougne, in the salinity range between 40 and 60‰ [9]. Beyond 60%, the oysters disappear from the environment. Too close to the ocean (3% higher than the salinity rate of the sea, in front of Djifère), the oysters are small in size (mode between 20 and 29 mm) with quite high numbers. A little inland, for salinities of the order of 40‰, we observe large oysters (mode between 40 and 49 mm) with a tighter distribution.

- *Medium-energy lagoon and biotope,* at the edge of the channels and in the mid-littoral and upper infra-littoral stages, on more or less muddy sand. The characteristic species of these environments are *Anadara senilis and Hemifusus morio*. We also find in this biotope *Conus papilionaceus* and *Murex cornutus* associated with *Zostera nana* herbariums. *M. cornutus* as well as *Cymbium pepo* are also found on sandbanks [10]. *Crassostrea gasar* is also observed in this biotope on hard bottoms at the edge of the mangrove vegetation.

Anadara senilis accepts all types of muddy bottoms with the exception of reducing muddy sediments with a high organic matter content [11]. The species is consumed in natural deposits by perforating mollusks such as *Thais forbesi, Thais callifera*, birds, crabs, and rays. The larger subjects are distributed further from the coast and the smaller ones within the salinity range of 35 to 55‰. The growth of the species is low, from 0.5 to 1 mm per year. It can live for 30 years, the age at which it reaches its maximum size, 75 mm. High mortality is observed in this species starting at 10 years of age.

- *A high-energy epi-marine biotope* where the lagoon population associated with many marine species can be found. *Natica fulminea* is harvested only in this environment [12]. This species is not accessible to female harvesters.

3.2 Sector and characterization of the actors in the exploitation of marine mollusks

The harvesting of marine mollusks (oyster, ark, cymbium) in the Saloum Delta mainly involves adult women who make up more than 95% of the stakeholders in this sector. Men of all ages and young girls represent the remaining fraction. This last group of actors is mainly present in the northern part of the delta, in the island villages of Dionewar and Bossinkang (98% of stakeholders).

The average age of these adult women is around 45 years. Their husbands work in a trade related to fishing.

Unlike other marine invertebrates, the collection of mangrove oysters is based on a clear sexual specialization. It is exclusively practiced by women. They keep full control of it from harvesting to processing and wholesale and retail sales.

The organization of the picking of the ark, the oyster, and the *Cymbium* is not the same. The harvesting of the ark is more structured. Women operate on foot or aboard pirogues propelled by the oar. As they do not have their own boats to travel to the collection areas, they resort to borrowing or renting. These canoes travel with two to three women. Outings limited to 1 day are alternated with the processing and drying of the collected products. On the other hand, the decision to collect cymbium is much more individual.

The collection of marine invertebrates is mainly the business of local people, no longer far from their villages. Migrants from Casamance, the Gambia, and Guinea Bissau are few in number (less than 1%). We meet them particularly in Missirah.

3.3 Marine invertebrate valuation methods and destination markets

The ways of valuing oysters have not evolved much. They are cooked or grilled. Cooking is done by boiling in water; the flesh is then removed from the open shells and lightly smoked and dried on racks. The shells are also opened on a wood fire, and the detached flesh is dried directly in the sun. Packaging and transport are done in bags or basins. Drying can take 2 to 3 days depending on whether the product is baked or smoked. Nearly 80% of oyster production is processed before being marketed.

The artisanal processing of ethmalose, the main species landed in the area, yields different products including "tambadiang" (salted and dried whole fish) and "metorah" (smoked and dried fish). Salt-dried (salted, fermented, and dried fish) and guedj (fermented and dried fish) are derived from other species.

The penetration of the monetary economy in this area has favored the strong marketing of oysters and other processed products. The distribution networks depend on the nature of the products. The fresh oysters are sold to tourist centers in the area, while the processed products are largely destined for the weekly markets in the surrounding areas (Passy, Sokone), large urban centers (Kaolack, Dakar), and the sub-region (Gambia). Salt-dried and metorah, little known to Senegalese consumers, are exported to Burkina Faso and Ghana.

3.4 Relative socio-economic importance for local communities

The very important fishery in the Saloum River delta provides annual catches of about 15,000 t, of which 350 are shrimp and 2600 are marine mollusks. The arch is the main resource exploited by the harvesting activity (1634 tons), followed by the murex, oyster, and cymbium, in that order. *Ethmalosa fimbriata* or cobo, the main species fished in the estuary, constitutes nearly 80% of the total landings of fish and crustaceans (shrimp) in the artisanal fishery.

These mangrove oyster collection activities are highly valued in terms of food, economy, and, to a lesser extent, culture.

Oysters contribute to the protein coverage of local populations because they are consumed fresh or processed. They can even be the staple food in some localities. They are used fresh or smoked to make the sauce to accompany white rice. Having a strong cultural value, oysters are consumed in large quantities during family ceremonies. Other artisanal processed products (guedj, tambadiang, yeet) are substitutes for fresh fish and condiments to flavor dishes.

As an additional remunerative activity during the dry season, fishing and marine invertebrate gathering activities are a brake on rural exodus. They are complementary to the winter agricultural activities of the area. They take place mainly from December to June. The harvesting of marine mollusks is mainly present in the island villages and in the continental Diomboss, constituting 29 and 52% of the fishermen, respectively. The fishing units involved in the harvesting of marine invertebrates are more important in the dry season than in winter (**Tables 1** and **2**). The vast majority of these fishing units seek arks and oysters in the dry season (1633 and 1499 fishing units out of a total of 3218). The "touffa" is only exploited in the dry season (1274 fishing units). Still in this section, we can notice the low number of fishing units that look for yeet (368 in the dry season, 350 in the rainy season).

Oyster farming occupies the first place in dry farming in the centers of Dionewar (189), Bétenti (175), and Bossinkang (145). This resource remains the main product sought by the harvesting fishing units in Dionewar (234 fishing units recorded) during the rainy season. The largest numbers of fishing units oriented to the harvesting of loincloth, tufa, and yeet are recorded in the dry season at Betenti, Bossinkang (minus the tufa), and Dionewar (minus the yeet). In wintering, there is only Dionewar for the loincloth in addition to the oyster.

The distribution by the geographical area of the number of women active in harvesting is shown in **Tables 3** and **4**.

The marine mollusk harvesting activity has seen a resurgence of interest over the past 10 years, with the entry of many women in the exploitation of this resource. In February 1999, some 2000 women were involved in shellfish exploitation in the Saloum Delta [13]. Young schoolchildren may accompany women collectors to the collection sites to earn some money to buy clothes and school supplies.

The shells are now systematically recovered in backfill and road works or calcined by open firing and crushed to provide lime with which bricks are made. These same powdered shells are also used to feed chickens. These new forms of valorization not only generate substantial incomes for the women but also stop the construction of shells piles, which used to occupy a lot of space and started to pose serious environmental problems.

Other no less important activities are associated with the mangrove environment present in the area. Thus, seaside tourism has particularly developed in the estuary of the Saloum Delta with the establishment of vacation villages, second homes, and campsites. This flourishing nature of the mangrove attracts many tourists who come to observe nature and migratory birds in their nesting sites and practice sport fishing and hunting. This activity is a source of employment for local communities and has greatly contributed to the development of local handicrafts. Tourism has also encouraged the construction of large pirogues for visitors in the Saloum bolongs.

AREA	SITE	TOTAL			AREA	SITE	TOTAL			AREA	SITE	TOTAL		
		Men	Women	TOTAL			Men	Women	TOTAL			Men	Women	TOTAL
SALOUM CONTINENT	BIL BAMBARA	1	0	**1**	**DIOMBOSS**	BAMBOUGAR MALICK	47	88	**135**		KATHIOR	17	6	**23**
	DIOBAYE	124	0	**124**		BAMBOUGAR MASSAMB	0	9	**9**		MAR SOULOU	5	20	**25**
	DJILOR DJIDIACK	28	0	**28**	**CONTINENT**	BANGALERE	43	0	**43**		MAR FAFACO	119	21	**140**
	FAOYE	93	0	**93**	*TOTAL*		*90*	*97*	***187***		MAR LOTHIE	19	58	**77**
	FATICK	45	0	**45**		BAKADADJI	2	17	**19**		MAYA	34	30	**64**
	FAYACO	297	0	**297**		BANI	16	0	**16**		MOUNDE	48	128	**176**
	FAYIL 1	28	0	**28**	**BANDIALA**	DASSILAME SERER	5	1	**6**		NDINDE	16	7	**23**
	FAYIL 2	38	0	**38**		MEDINA SANGAKO	62	34	**96**		NGADIOR	79	71	**150**
	FELIR	70	0	**70**	**CONTINENT**	MISSIRAH	374	18	**392**		NIODIOR	76	52	**128**
	FIMELA	115	0	**115**		NEMABA	55	21	**76**	*TOTAL*		*2472*	*2091*	***4563***
	FOUNDIOUGNE	195	0	**195**	*TOTAL*		*514*	*91*	***605***	**TOTAL**		**4842**	**2285**	**7127**
	GAGUE CHERIF	82	0	**82**	ISLAND VILLAGES	BAKHALOU	48	6	**54**					
	GAGUE MODY	2	0	**2**		BAKHOUSS	3	2	**5**					
	GAMBOUL	46	0	**46**		BAOUT	85	54	**139**					
	KAHONE	10	0	**10**		BASSAR	243	56	**299**					
	KAOLACK	16	0	**16**		BASSOUL	82	39	**121**					
	KOYLAL	15	0	**15**		BETENTI	605	565	**1170**					
	LATMINGUE	32	0	**32**		BOSSINKANG	178	253	**431**					
	NDAKHONGA	28	0	**28**		DIAMNIADIO	126	66	**192**					

DOI: http://dx.doi.org/10.5772/intechopen.109415

AREA	SITE	TOTAL			AREA	SITE	TOTAL			AREA	SITE	TOTAL		
		Men	**Women**	**TOTAL**			**Men**	**Women**	**TOTAL**			**Men**	**Women**	**TOTAL**
	NDANGANE DIENE	64	0	**64**	DIOFANDOR	26	14	**40**						
	NDANGANE SAMBCU	343	6	**349**	DIOGANE	57	85	**142**						
	NDIAFFATE TOUCOUL	11	0	**11**	DIONEWAR	296	227	**523**						
	NDOLETTE	50	0	**50**	DIOUGNDJ	9	13	**22**						
	NERAN	27	0	**27**	DJINACK	50	81	**131**						
	NIAMDIAROKH	6	0	**6**	DJIRNDA	89	75	**164**						
TOTAL		*1766*	*6*	*1772*		FALIA	65	88	**153**					
						FAMBINE	77	70	**147**					
						GOUK	20	4	**24**					

Table 1.
Distribution by survey site and area of the number of fishers (by sex) in the Sine-Saloum region during the dry season.

AREA	SITE	TOTAL		All genders	ZONE	SITE	TOTAL		All genders	ZONE	SITE	TOTAL		All genders
		Men	Women				Men	Women				Men	Women	
	BIL BAMBARA PEUL	22	0	**22**		BAMBOUGAR EL HADJ	52	0	**52**		FALIA	48	101	**149**
	CADDO (B. BOLONG)	12	0	**12**	**DIOMBOSS**	BAMBOUGAR MALICK	81	0	**81**		FAMBINE	57	22	79
	DIOBAYE	78	0	**78**		BAMBOUGAR MASSAMBA	52	0	**52**		FELIR	133	0	**133**
	DJILOR DJIDIAK	42	34	**76**	**CONTINENT**	BAMBOUGAR MOMATH	12	0	**12**		GOUK	10	2	**12**
	FAOYE	200	0	**200**		BANGALERE	24	0	**24**	**ISLAND VILLAGES**	KATHIOR	10	8	**18**
	FATICK	103	0	**103**		LERANE COLY	18	0	**18**		KOULOUK	5	2	7
	FAYACO	151	0	**151**		SOKONE	55	0	**55**		MAR FAFACO	82	34	**116**
	FAYIL 1	94	0	**94**	*TOTAL*		294	0	***294***		MAR LOTHIE	47	65	**112**
	FIMELA	38	0	**38**		BAKADADJI	20	0	**20**		MAR SOULOU	4	45	**49**
	FOUNDIOUGNE	298	0	**298**		BANI	25	0	**25**		MAYA	70	7	77
	GAGUE CHERIF	79	8	**87**		DASSILAME	22	0	**22**		MOUNDE	51	93	**144**
	GAMBOUL	24	0	**24**	**BANDIALA**	MEDINA	116	0	**116**		NDINDE	9	11	**20**
SALOUM	KAHONE	21	0	**21**		MEDINA SANGAKO	7	0	7		NGADIOR	68	21	**89**
	KAMATANE	17	0	**17**	**CONTINENT**	MISSIRAH	327	3	**330**		NIADIARA	2	3	**5**
CONTINENT	KAOLACK	23	0	**23**		NEMABA	59	0	**59**		NIODIOR	319	4	**323**
	KOYLAL	10	0	**10**		SOUKOUTA	8	0	**8**	*TOTAL*		2739	1268	***4007***
	LATMINGUE	7	0	7		TOUBACOUTA	97	0	**97**	**TOTAL**		**6269**	**1319**	7588
	NDAKHONGA	45	0	**45**	*TOTAL*		681	3	***684***					
	NDANGANE SAMBOU	126	6	**132**		BAKHALOU	6	4	**10**					

AREA	SITE	TOTAL		All genders	ZONE	SITE	TOTAL		All genders	ZONE	SITE	TOTAL		All genders
		Men	Women				Men	Women				Men	Women	
	NDIAFATTE	57	0	**57**		BAKHOUSS	10	3	**13**					
	NDOLETTE	83	0	**83**		BAOUT	159	65	**224**					
	NIAMDIAROKH	1	0	**1**		BASSAR	105	45	**150**					
	PALMARIN DIAKHANOR	66	0	**66**	**ISLAND VILLAGES**	BASSOUL	48	18	**66**					
	PALMARIN NGALLOU	358	0	**358**		BETENTI	444	55	**499**					
	PALMARIN NGOUDOUMANE	32	0	**32**		BOSSINKANG	199	249	**448**					
	ROKH	38	0	**38**		DIAMNIADIO	243	58	**301**					
	SANDIKOLI	41	0	**41**		DIOFANDOR	31	5	**36**					
	SASSARA	116	0	**116**		DIOGANE	84	29	**113**					
	SIBASSOR	21	0	**21**		DIOGAYE	35	4	**39**					
	SOUM	54	0	**54**		DIONEWAR	204	226	**430**					
	THIANGANE	54	0	**54**		DJINACK BARA	71	29	**100**					
	TOURNAL NONANE	200	0	**200**		DJINACK DIATAKO	16	28	**44**					
	VELOR KEUR DE DEMBA	44	0	**44**		DJIRNDA	169	32	**201**					
TOTAL		2555	48	*2603*										

Table 2.
Distribution by survey site and environment of the number of fishers (by sex) in the Sine-Saloum region during the rainy season.

	Fishing			Harvesting			TOTAL		TOTAL
Areas	**Men**	**Womens**	**total**	**Men**	**Womens**	**total**	**hommes**	**femmes**	
SALOUM CONTINENT	1766		1766		6		1766	6	1772
DIOMBOSS CONTINENT	90		90		97		90	97	187
BANDIALA CONTINENT	510		510	4	91		514	91	605
ISLAND VILLAGES	2453		2453	19	2091		2472	2091	4563
TOTAL	4819	0	4819	23	2285	0	4842	2285	7127

Table 3.
Distribution by gender and geographical area of the number of people active in fishing and gathering in the Sine-Saloum region during the dry season.

	Fishing			Harvesting			TOTAL		TOTAL
Areas	**Men**	**Women**	**total**	**Men**	**Women**	**total**	**Men**	**Women**	
SALOUM CONTINENT	2546		2546	9	48	57	2555	48	2603
DIOMBOSS CONTINENT	254		254	40		40	294	0	294
BANDIALA CONTINENT	666		666	15	3	18	681	3	684
VILLAGES INSULAIRES	2635		2635	104	1268	1372	2739	1268	4007
TOTAL	6101	0	6101	168	1319	1487	6269	1319	7588

Table 4.
Distribution by gender and geographical area of the number of people active in fishing and gathering in the Sine-Saloum region during the rainy season.

Apiculture is a significant activity in mangrove areas. The collection of honey contributes to the improvement of rural household incomes. The exploitation is free for the indigenous populations and is mainly carried out by men.

The mangrove provides other no-less-important products. Local villagers cut firewood and make charcoal. Mangrove wood is widely used for fish smoking. This resource also provides poles to support houses and very resistant construction wood.

The information received during the semi-structured interviews highlighted the medicinal (leaves, bark), religious (sacred forests), and esthetic (necklaces made from small oyster shells) uses of the mangrove.

3.5 Price level and remuneration of the actors

Products that are not very perishable, are easy to sell, and processed by hand are in high demand on the market and at sufficiently remunerative prices.

As detailed information is not available for all the products listed, the oyster industry is used as an illustration. Thus, we note that for dried oysters, the price per kilo is 1000 to 1500 FCFA in the production centers, 2000 at the weekly markets, and 2500 FCFA in the large urban centers. The oyster lids are exported to Asian countries. The price per kilo is 40,000 FCFA at the level of the processing center. The shells are recycled and contribute more than 10% of the women's income.

Investments in oyster harvesting activities are very basic and are limited to rudimentary instruments consisting of a cutter and a few baskets made by the women themselves.

Operating expenses are virtually nil. The dugout canoes used are propelled by oars. Women without boats rent or borrow them. Many of them use the same pirogue to make their jobs profitable. Payment is usually in kind, such as a basket of smoked oysters at the end of the season.

For a total production of 5 tons of processed products (335 tons fresh equivalent), the commercial value of oysters collected and sold in Saloum is estimated at 30 million FCFA [14].

3.6 Causes and impacts of mangrove deforestation on the estuarine ecosystem

Surveys conducted by CRODT from April to December 2000 indicate income levels between 10,000 and 50,000 CFA francs per season, with a maximum of 140,000 francs for some women.

The income generated is generally earmarked for women's personal needs, domestic needs, family ceremonies, and childcare (clothing and schooling). Some women save their income for an entire season for an event planned well in advance. Very few sustainable investments are made from these earnings although they provide women with some financial independence.

For all fishery products taken together, the added value generated by fishing activities in this estuarine environment has been estimated at 2 billion FCFA [15].

There are no data on the real importance of mangroves in the different ecological zones of the Saloum Delta that would allow to establish a reference state to serve as a comparison of the action of factors that could lead to the disappearance of mangroves. The water balance for the period 1961–1990 has enabled various researchers to identify the factors responsible for the disappearance of the Sine-Saloum mangrove [16–18]. These include: (i) the decrease in freshwater inputs related to salinization groundwater contamination; (ii) upwelling of the tidal boundary upstream; (iii) contamination of the water column; (iv) the increase in mangrove mortality (100% recorded in Foundiougne); (v) the disappearance of *Avicennia* and the appearance of live tannins; and (vi) the resurgence of the phenomena of deposition and eolian reworking.

The periods of drought of the past three decades and the multiform aggressions of the populations on the environment have thus profoundly modified the hydrological and biogeochemical functioning of the mangrove ecosystem of Sine-Saloum and significantly degraded the stocks of marine mollusks. As a result, the age-old balances that have always existed between the populations and their environment have been disrupted. Thus, at the environmental level, hyper-salinity and acidification of most soils and consequently the reduction of biodiversity [19–21] have been observed throughout the Sine-Saloum catchment area, thus threatening shrimp fisheries. At the socio-economic level, the disturbances generated by these series of degradation have led to a drop in food production (agricultural, fishery, and forestry) and in the income of the populations.

In addition, there has been a decrease in the size of harvested individuals [22]. As the deposits are depleted in the immediate surroundings of the villages, women are forced to go far from their land in search of new sites. This leads to an increase in production costs, a resurgence of conflicts, and an increase in poverty, which are already accentuated in the Saloum Islands.

Beyond the fishing and gathering activities themselves, the disappearance of the Sine-Saloum mangrove has created various problems at the ecosystem level: (i) Difficulty of navigation in certain parts of the estuary and its secondary branches. Today, one can no longer access villages like Niodior or Dionewar at low tide. (ii) Narrowing of the niominkas' agricultural space due to the progression of tannins and the decrease in water reserves. Coastal erosion is considerably reducing arable land. (iii) Displacement of populations to safer sites (case of Palmarin after the heavy storms of 1987). Villages such as Djifère, Ngallou, and Diakhanor are also threatened by coastal erosion. (iv) Threat to tourist activity—chronic food deficit of nearly 47% of primary needs for wood for energy and drinking water.

4. Recommendations

The exponential growth of the population (3%) and its needs, as well as the modification of the marine environment (rupture of the Sangomar river among others), constitutes, on the one hand, a serious threat to the preservation of the mangrove's malacological resources and, on the other hand, a hindrance to the socio-economic development of the area because of the risks of food insecurity.

To reverse the situation, a number of management measures have been recommended, including the institution of biological rest and rotation in the collection of marine invertebrates, the restoration of the ecosystem, and the establishment of an institutional framework to ensure the sustainable management of renewable resources.

4.1 Biological recovery and rotation in collection

Biological recovery, in the true sense of the word, aims to protect juveniles and to ensure a spawning season without anthropogenic disturbance. It is to be distinguished from a reduction in fishing (or harvesting) effort over the period of the year. The establishment of a biological rest contributes to ensuring the sustainability of the resource. It is well known that traditionally, the oyster benefits from a period during which harvesting is not practiced. This period generally corresponds to the rainy season (from June to October). The measure has always been adopted, not for reasons of resource protection but rather for reasons of product quality, because sexually mature oysters are said to be milky and tasteless.

Different forums organized with the actors of marine mollusk collection have allowed the mapping of the fishing areas and the consideration of management means for the exploited resources. Rotation systems were defined for the marine mollusk collection sites (oyster, ark, murex, cymbium) in the RBDS area with a biological recovery period of 5 months (June to October), a selective harvesting period (October to January), and an access period (January to June).

The closure of harvesting areas also leads to better growth of marine mollusks and preservation of the resource with a view to ensuring significant and sustainable incomes for harvesters.

Thus, the stakeholders of oyster harvesting in the Sine-Saloum Delta have accepted the development of harvesting areas for this resource in order to improve its productivity. Beach committees were created in some villages, and shellfish harvesting sites were closed in 2001. The impact of these measures has yet to be evaluated.

Beyond the oyster, which represents the most commercially interesting resource (even other products have very high prices per kilogram), it will be necessary to consider the establishment of biological resting on the arches. This type of resource is exploited all year round at the ecosystem level. The commercial and alimentary importance of arches, their permanent availability, means that they must be the subject of a specific scientific monitoring activity that should lead to the establishment of a plan for the exploitation of the different deposits taking into account the bio-ecological knowledge of the species.

4.2 Implementation of sustainable marine invertebrate harvesting practices

The marine mollusk harvesting pressure exerted by Missirah women on oysters is estimated at 23.7% of the available stock (22). This level of exploitation is related to the oyster harvesting methods practiced by the women of this locality: (i) selective harvesting of the largest individuals; (ii) no systematic cutting of mangrove roots; (iii) only one annual passage per harvest site.

Such practices need to be systematized throughout the Sine-Saloum Delta to ensure the sustainable management of natural resources.

4.3 Restoration and rehabilitation of ecosystems

The rehabilitation of degraded habitats and the restoration of ecosystems are becoming over time increasingly important tools for preserving biodiversity and ensuring the sustainability of exploited resources. Rehabilitation aimed at restoring essential ecological functions in modified environments (by man or nature) is particularly appropriate in drought-prone areas.

More ambitious than rehabilitation, restoration aims to restore degraded environments (by man and/or nature) to their natural condition. However, since it is difficult to determine the initial condition of ecosystems and since ecosystems are constantly changing, restoration is rarely a realistic goal. Several authors argue that restoration of ecosystems do not always require intervention because, left in their natural condition, many ecosystems return to a condition close to "pre-disturbance."

Specific mangrove reforestation actions have already taken place with very poor results, particularly in the Missirah area, under the leadership of NGOs and local youth associations. The ecological particularity of the brackish wetland of the mangrove areas added to the complexity of the reproduction and implantation of young propagules in the mud require that studies must be carried out in parallel with the field actions in order to determine the best conditions for the practice of mangrove reforestation.

4.4 Establishment of an institutional framework for the rational management of environmental resources

The management of fisheries, in terms of both the development of laws and regulations and their application, control, and monitoring, is a process that, until recently, was the exclusive responsibility of public authorities. In recent years, factors related to the scarcity of the resource, the scientific insufficiency of the resources (potentialities, catches, etc.), the resurgence of conflicts, the emergence of professional organizations, as well as the resource management problems faced by the fisheries administration have highlighted the need for concerted management of renewable resources. This approach requires the establishment of an institutional framework bringing together research, administration, NGOs, and professional organizations.

Several institutions including the DPN, DPCA, CRODT, ITA, IUCN, and FOs can be involved in this process.

The DPCA is the institutional framework in charge of putting in place the conditions to achieve the major goals assigned to inland fisheries, including the development and preservation of resources.

The task of CRODT is to monitor the evolution of the fisheries and stocks and to prepare fishery management plans that take into account both biological factors and socio-economic constraints.

The Institute of Food Technology (ITA in French) is in charge of research on the improvement of processing and packaging techniques for fishery products.

The World Conservation Union (IUCN), a leader in the management of so-called “wild” resources, can assist in the formulation of appropriate management policies.

The National Parks Department (DPN in French) manages a large part of the Saloum estuary environment and pilots several projects for the development and management of natural resources in various fields (forestry, fishing, hunting, tourism, etc.) with various national and foreign partners.

Many professional organizations have emerged in the past 10 years. The CNPS and FENAGIE-Pêche are the most representative of them.

5. Conclusions

The mangrove case study provides a better understanding of the problem of the use of malacological resources in terms of the economic benefits that can be derived from them as well as in terms of the sustainability of use patterns by determining the economic and social values that the populations of the Sine-Saloum Delta attach to these resources. The assessment of the impacts of deforestation on the sustainability of the exploitation of malacological resources led to recommendations relating to the establishment of biological resting areas, the institution of healthy practices for the collection of marine invertebrates, the restoration and rehabilitation of ecosystems, as well as the establishment of an institutional framework for the sustainable management of natural resources in the estuary complex of the Saloum Delta Biosphere Reserve.

Acknowledgements

This work was funded as part of the IUCN Senegal Values Project on Wild Forest and Fisheries Resources, the research component of which was entrusted to the Senegalese Institute for Agricultural Research ISRA. We thank them for that. We would also like to thank Dr. Bocar Sabaly BALDE for his support in the English translation of this document.

Conflict of interest

The authors declare no conflict of interest.

Appendices and nomenclature

CNPS	National Collective of Artisanal Fishermen of Senegal
CRODT	Oceanographic Research Center of Dakar-Thiaroye
DPCA	Department of Continental Fisheries and Aquaculture
DPN	Department of National Parks
FENAGIE-Pêche	National Federation of Fishing EIGs
ITA	Institute of Food Technology
PO	Professional Organizations
RBDS	Saloum Delta Biosphere Reserve
IUCN	World Conservation Union
VALUES	Valorization of species for sustainable use of sustainable resources in Senegal

Author details

Hamet Diaw Diadhiou[1*], Moustapha Deme[2] and Djiby Thiam[3†]

1 Fishery Biologist, Retired researcher from the Senegal Institute of Agricultural Research, Senegal

2 Economist researcher at the Senegal Institute of Agricultural Research, Senegal

3 Former Researcher at the Senegal Institute of Agricultural Research, deceased Former, Senegal

*Address all correspondence to: hamet_diadhiou@yahoo.fr

† Deceased

References

[1] Dacosta H. Variabilité des précipitations sur le bassin versant du Salaam IN: Gestion des ressources côtières et littorales du Sénégal. Diaw AT, Ba A, Bovlond P, Diouf PS, Lake LA, Mbow MA, et al, editors. Actes de l'atelier de Gorée 27-29 juillet. 1992. pp. 87-103

[2] Ba M, Barusseau JP, Demarcq C, Descamps C, Diop S, Diouf B, et al. Evolution côtière au quaternaire récent. IUGS/UNESCO/COMARAF/ASEQUA/EPEEC/ORSTOM/UCAD/Université de Perpignan. 1993. p. 32

[3] EPEEC. Rapport final de l'atelier d'étude des mangroves et de l'estuaire du Saloum (Sénégal). UNESCO/EPEEC, Dakar (Sénégal). 1982. p. 175

[4] MEPN. Stratégie nationale et plan National d'Actions pour la conservation de la biodiversité. 1998. p. 94

[5] Diouf PS. Les peuplements de poissons des milieux estuariens de l'Afrique de l'Ouest: l'exemple de l'estuaire hypersalin du Sine-Saloum. [thèse de Doctorat], Univ. De Montpellier II, 1996. p. 267

[6] Diouf PS, Kebe M, Le Reste L, Bousso T, Diadhiou H, Gaye AB. Plan d'action forestier: pêche et aquaculture continentales, CRODT/MDRH/FAO, Diagnostic. 1991: Vol. 1, p. 263

[7] Barusseau JP, Diop ES, Giresse P, Monteillet J, Saos J-L. Conséquences sédimentologiques de l'évolution climatique fini-holocène (10 puissance 2 - 10 puissance 3) dans le delta du Saloum (Sénégal). Océanographie Tropicale. 1986;**21**(1):89-98 ISSN 0245-9418.

[8] Marazova AL, Leung-Tack KD, Kholodov VI, Trousevich VV, Camara S, Maskevski VK, et al. L'ostréiculture en milieux de mangroves (Etudes de cas en Guinée et au Sénégal). Comaraf. 1991;7: 148

[9] Leung-Tack TD, Vincke PP. Une étude de la population de l'huître de palétuvier *Crassostrea gasar* (Adanson), dans le Saloum, de Djifère à Foundiougne. Etude des mangroves et estuaires du Sénégal: Saloum et Somone, EPEEC, Rapport final. Dakar. 1984. pp. 31-46

[10] Elouard P, Rosso JC. Biogéographie et habitat des mollusques actuels du delta du Saloum (Sénégal). Géobios. 1977; **10**(2):275-299

[11] Zabi SF, Le Loeuff P. Revue des connaissances sur la faune benthique des milieux marginaux littoraux d'Afrique de l'Ouest. Rev. Hydrob. Trop. 1992; **25**(3):209-251

[12] Dème M, Diadhiou HD, Thiam D. Recensement des unités de pêche dans les zones géographiques du Fleuve Sénégal et du Complexe deltaïque du Sine-Saloum en 1999. 2000: Document technique du projet UDRSS-VALEURS. p. 131

[13] Dème M, Diadhiou HD, Thiam D. Effort de pêche, captures spécifiques et valeurs économiques de la pêche continentale dans le fleuve Sénégal et au Sine Saloum. 2001: Rapport UICN/Projet Valeurs, + planches. p. 30

[14] Dème M, Diadhiou HD, Ndiaye V, Thiam D. Synthèse des résultats d'enquêtes du Projet Valeurs et capitalisation des données d'inventaire sur les ressources sauvages au Sénégal. Document ISRA CRODT, CSE. 2005. p. 24

[15] EPEEC. L'atelier d'étude des mangroves au Sud de l'estuaire du Saloum : Diombass, Bandiala (Sénégal): Rapport technique UNESCO, Division des Sciences de la Mer, Rosta, Dakar. 1983. p. 219

[16] EPEEC. Etude des mangroves et estuaires du Sénégal: Saloum et Somone. UNESCO/Division des Sciences de la Mer: Rapport final EPEEC. 1984. p. 32

[17] Diatta L, Bodian A, Thoen D. Etude phyto sociologique: Rapport final de l'atelier d'étude des mangroves et de l'estuaire du Saloum (Sénégal). 1981. pp. 121-157

[18] Diop ES. La côte ouest-africaine du Sénégal à la Mellacorde (Rép.Guinée): Editions ORSTOM, Collection Etudes et [thèses]. 1990. p. 379

[19] Sadio S. Pédogenèse et potentialités forestières des sols sulfates acides des tannes du Sine [thesis]. Wageningen University: Landbouwetenschaffen; 1991

[20] Diouf PS, Thiam D, Sène C, Dia A, Ly ME, N'Diaye NA. Aménagement participatif des pêcheries artisanales du Sine-Saloum (Sénégal): Rapport EPEEC. 1998. p. 50

[21] Deschamps C. La collecte des arches, une activité bimillénaire dans le Bas-Saloum (Sénégal). In: Dynamique et usages de la mangrove dans les pays des rivières du sud (du Sénégal à la Sierre Léone), Cormier-Salem MC, editors. Colloques et séminaires, Editions ORSTOM. 1994. pp. 107-113

[22] Bay T. Contribution à l'évaluation du stock d'huîtres de palétuviers (Crassostrea gasar ADANSON) dans le parc national du Delta du Saloum (Sénégal): Mémoire de fin d'études en vue d'obtenir le grade d'Ingénieur Agronome, orientation Elevage. Faculté des Sciences Agronomiques de GEMBLOUX, Belgique + Bibliographie + Annexes. 2000. p. 75

Chapter 5

Mangrove Health Assessment Using Hemispherical Photography: A Case Study on Mangrove Ecosystem for Ecotourism at Tajungan-Bangkalan, Madura Island, Indonesia

Maulinna Kusumo Wardhani

Abstract

Mangrove health status indicates sustainable management and efforts to control forest damage. The hemispherical photographic method facilitates the observation and monitoring of forest health. This method is also more accessible, faster, and practical than the conventional method. Data analysis in this method requires Image J software. The selection of research sites aims to determine forest management for mangrove ecotourism. Thus, rehabilitation decision-making is right on target. The research results on the health status of mangroves in Tajungan-Bangkalan, Madura Island, Indonesia, showed moderate-to-good levels. The percentage of cover in the good category is at the MDRT01 station, which is 81.64 ± 3.35%. The MDRT02 observation station has a closing percentage of 64.31% ± 20.41% and is in the moderate category. The suitability of tourism based on the percentage of closure is in the appropriate category at the MDRT02 station and very suitable at the MDRT01 station. The results of this study could be used for planning of mangrove ecotourism and also for education on planting of mangrove seedlings.

Keywords: mangrove, health, hemispherical photography, ecotourism, mangrove ecosystem

1. Introduction

Mangrove forest is a coastal ecosystem with a critical ecological role in the shallow water zone. This collection of vegetation forms an ecosystem with environmental services producing nutrients for aquatic organisms, helping the carbon cycle, and protecting organisms and the coastal environment. In addition, mangrove ecosystems have benefits that are currently developing as conservation, rehabilitation, and educational areas to increase community welfare [1–3].

Coastal development, expansion of aquaculture, overfishing, and climate change threaten the existence of mangroves worldwide. As a result, the mangrove forest area decreased by 62% between 2000 and 2016 [4]. Research into the causes of mangrove loss over the last 20 years reveals that socioeconomic and biophysical factors account for most of the degradation, despite increasing mangrove cover in some areas [5].

A decline in the quantity and quality of mangrove forests has also occurred in Indonesia, particularly in Tajungan Village, Kamal District, Bangkalan Regency on Madura Island. The mangrove ecosystem in Tajungan Village is an area that controls coastal conditions from the threat of abrasion, land subsidence, and seawater intrusion. Cultivated land, housing, and others have changed the mangrove area through logging/deforestation. Preliminary observations in the field noted problems in managing mangrove ecosystems in Tajungan Village, Kamal District, Bangkalan Regency, based on environmental conditions, including tree felling, beach abrasion, garbage, and damage caused by local communities. This problem causes a decrease in the environmental quality of the mangrove ecosystem in this area. In addition, the technical problems in managing this mangrove ecosystem are the limited human resources in terms of numbers, education, management experience, service, and supervision.

One of the efforts to utilize and preserve mangrove ecosystems is through ecotourism activities. Appropriate and environment-based planning and management of coastal areas are necessary to protect the sustainability of mangrove ecosystems from achieving sustainable regional development [6, 7]. One of the efforts made by the village of Tajungan is the development of mangrove ecotourism as a conservation and educational effort.

The basic principle of ecotourism is to enhance conservation. However, few studies have assessed their effectiveness in meeting conservation objectives and whether the type of tourism activity affects outcomes. Small-scale tourist sites have more considerable social conservation outcomes, including conservation ethics, perceptions, attitudes, and behavior changes. The type of tourism, and the associated incentives, can significantly affect conservation outcomes [8]. The coastal ecotourism that is currently developing is mangrove ecotourism. One of the best opportunities is to ensure that mangrove-based tourism is carried out based on sustainable principles [9]. One of the criteria for due diligence on mangrove ecotourism is the percentage of forest cover [10, 11]. This closure percentage can also indicate the health of the mangrove ecosystem [12]. Using the hemispherical photography method, the mangrove health assessment technique based on the percentage of mangrove cover produces more accurate data with easy application. This technique uses a camera with a viewing angle of 180 degrees at one point of capture [13, 14]. The definition of canopy cover percentage is the vertically projected portion of the land surface that is overgrown with plants [13]. The result is the percentage of community canopy cover, which is one of the main components of the Mangrove Health Index (MHI) [15]. Current research rarely discusses mangrove ecotourism in Indonesia and its relation to health status. Therefore, this study aims to assess the health status of mangroves based on the percentage canopy cover and their suitability for tourism areas. The result of this research helps ensure the preservation of the mangrove ecosystem as a tourism resource in Tajungan Village.

2. Method

2.1 Stations and observation plots

This research was conducted in Tajungan Village, Kamal District, Bangkalan Regency, East Java, Indonesia. The research locations are presented in **Figure 1** and **Table 1**.

2.2 Research methods

2.2.1 Identification of potential station points

The step to determine the observation station (To) begins with the interpretation of mangrove objects, namely identifying the distribution of mangroves in regional stations using Google Earth which provides an initial spatial picture of the existence of

Figure 1.
Station and observation plots.

No	Station/plot	Coordinate
1	**MDRT01**	
	MDRT01_01	S 07.15586E 112.69589
	MDRT01_02	S 07.15592E 112.69598
	MDRT01_03	S 07.15596E 112.69603
2	**MDRT02**	
	MDRT02_01	S 07.15584E 112.69600
	MDRT02_02	S 07.15592E 112.69598

Table 1.
Station locations and observation plots.

mangrove ecosystems. The determination of the number of stations takes into account regional representation, time availability, resources, and budget. Subsequently, potential stations were created as candidates for permanent monitoring sites and required verification for inclusion on a provisional thematic map. After the establishment of permanent monitoring plots, species identification was carried out based on Tomlinson's [16] reference. If there are doubts about the identification, the researcher takes photos of the parts of the mangrove plant, namely stands, roots, stems, leaves, flowers, fruit, and samples, for further identification in the laboratory with the help of literature or the help of mangrove identification experts. Researchers must record all data obtained using worksheets on waterproof paper.

2.2.2 Data collection

This study used the line transect method by making the plots perpendicular to the coast toward the land. Placement of plots is by using stratified random sampling in each stratum by considering the ease of access to the observation sites.

The number of research stations is two, with each station consisting of three and two2 plots, so there are five plots in this study. Observation plots were made parallel to the coastline, measuring $10 \times 10\ m^2$ using a rolling meter and surrounded by rope.

2.2.3 Cover percentage

Collecting mangrove cover percentage data follows the steps of collecting mangrove community data. Analysis of the percentage of mangrove cover uses the hemispherical photography method, which requires a camera with a viewing angle of 180 degrees at one shooting point [13, 14]. This technique tends to be relatively new in Indonesia with its application to mangrove forests. The photos taken in this study used a 24 MP smartphone front camera with 1:1 frame mode. However, implementing this method is very easy and produces more accurate data. The steps for implementing this method are that each plot measuring $10 \times 10\ m^2$ is divided into four (four) quadrants measuring $5 \times 5\ m^2$. The firing points are around the center of the small square; they had to get between one tree and another and avoid shooting right under the tree trunk. The position of the camera is parallel to the chest height of the researcher/team who took the photo and is perpendicular/facing straight to the sky. Photo numbers are recorded on a data sheet form to simplify and speed up data analysis. Minimum shooting is done at four quadrant points with each plot measuring $10 \times 10\ m^2$, without repeating and marking photos at the end of each shooting session in each plot. Taking photos in this study at least 20 points with the assumption that there are five plots, and each plot has four quadrants of shooting points. When shooting avoid taking multiple photos at each point to prevent confusion in data analysis [17]. **Figures 2** and **3** present illustrations of taking photos of mangrove cover.

2.3 Data analysis

The analysis of percentage cover data was done using Image J software by separating sky pixels and vegetation cover. Converting a photo to 8 bits is the first step to separate the canopy and sky into a single color line, namely from white (0) to black (255). Next, calculate the number of pixels of the sky (white) and canopy (black) in the histogram menu of the Image J software. Finally, the percentage of canopy cover is

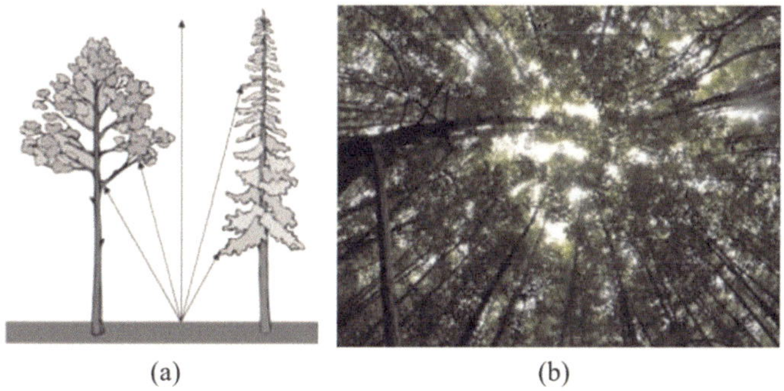

Figure 2.
(a) Illustration of the hemispherical photography method for measuring mangrove cover [13, 14]; (b) the results of shooting using a fisheye lens vertically [17].

Figure 3.
Shooting points in each observation plot [17].

the ratio of the number of pixels of the canopy (P_{255}) divided by the total number of pixels (P_{tot}) multiplied by 100% for each observation photo (Eq. (2)):

$$P_{255} = P_{\text{tot}} - P_0 \quad (1)$$

$$C = \frac{P_{255}}{P_{tot}} \times 100\% \quad (2)$$

Information:
C = percentage of canopy cover (%)
P_{255} = the number of pixels with a value of 255 (canopy).
P_0 = the number of pixels with a value of 0 (sky)
P_{tot} = the total number of photo pixels

2.4 Interpretation of results and determination of mangrove community conditions

The analysis results will produce a density value in units of trees/ha and the percentage of cover in percent units (%). These results can describe the status of the condition of mangrove forests which are categorized based on the Decree of the Minister of Environment No. 201 of 2004, as presented in **Table 2**.

3. Results and discussion

Observation of the mangrove community in Tajungan Village, Kamal District, Bangkalan Regency, East Java, was carried out at two locations designated as stations,

Criteria	Cover (%)
Good	≥75
Moderate	50–75
Damage	<50

Table 2.
The standard for damage to mangrove forests based on the decree of the minister of environment No. 201 of 2004.

No	Station	Species
1	MDRT01	*Avecennia marina* *Rhizopora apiculata* *Rhizopora mucronata* *Sonneratia alba*
2	MDRT02	*Avecennia marina* *Avecennia alba* *Avecennia officinalis* *Avicennia rumphiana* *Sonneratia alba*

Table 3.
The species of mangrove species at each station.

namely MDRT 01 and MDRT 02. The mangrove forest ecosystem in the study area consisted of seven species spread across observation stations, namely *Avicennia marina*, *Avicennia alba*, *Avicennia officinalis*, *Avicennia rumphiana*, *Sonneratia alba*, *Rhizophora mucronata*, and *Rhizophora apiculata* (**Table 3**). The mangrove forests in the study area have varied types and conditions with a slightly sandy mud substrate. In addition, solid anthropogenic waste covers 40% of the observed area of the mangrove ecosystem at this location.

The process of analyzing the percentage of mangrove forest cover is shown in **Figure 4**. After separating the photo into two colors, calculate the number of pixels using the histogram menu on the analysis tab (**Figure 5**). This mode displays the number of black-and-white pixels.

The results of taking photos of the canopy cover using the hemispherical photography method and analysis using the Image J software structure in Tajungan Village,

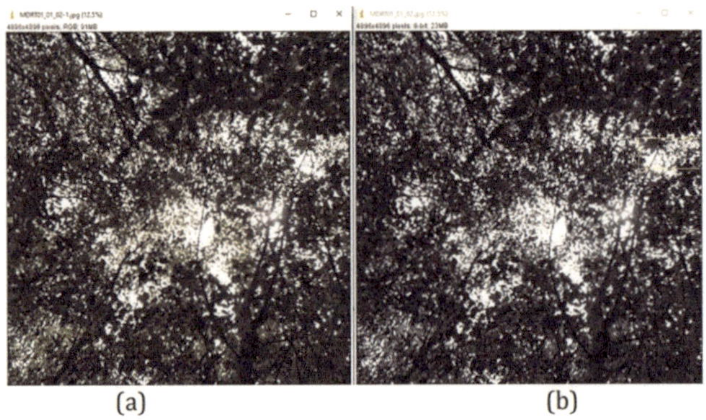

Figure 4.
(a) Photos opened using image-J, and (b) photos that have been converted to 8-bit image.

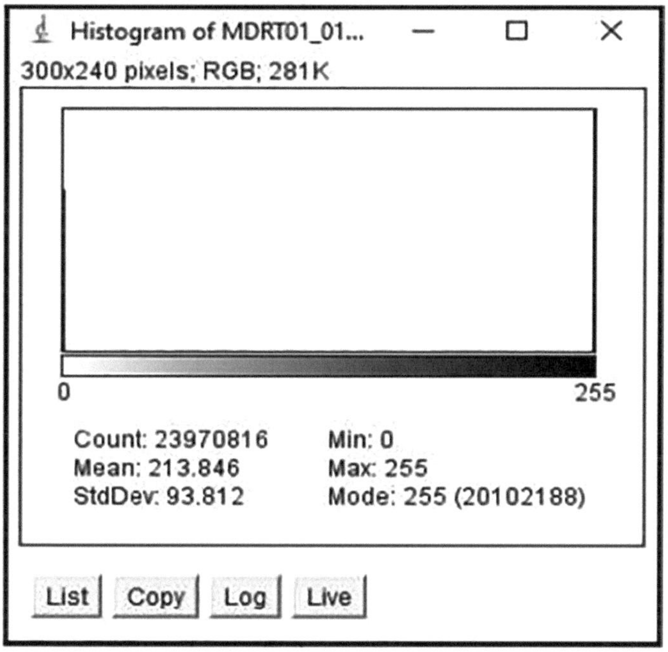

Figure 5.
The histogram shows that the analyzed photos have almost 24 million pixels (count), dominated by canopy pixels (255) with 20,102,188 black pixels.

Kamal District, Bangkalan Regency, East Java, are presented in the following. The percentage of canopy closure for each station is the average of all plots in **Table 4**. The percentage of canopy closure by category of mangrove damage is presented in **Table 5**.

The analysis results show that the highest percentage of canopy cover is at station MDRT01 (81.64%) and station MDRT02 (64.31%). The status of mangrove damage refers to Minister of Environment Decree Number 201 of 2004 concerning standard

Quadrant	Cover canopy (%)		
	Plot 1	Plot 2	Plot 3
	MDRT01		
1	76.16	78.65	83.56
2	86.95	66.75	82.26
3	82.76	81.29	87.05
4	79.12	73.62	84.39
5	84.57	80.83	86.99
6		78.51	
7		88.43	
8		75.55	
9		79.80	
Average	**81.91**	**78.16**	**84.85**

Quadrant	Cover canopy (%)		
	Plot 1	**Plot 2**	**Plot 3**
	MDRT02		
1	16.65	85.96	
2	32.12	76.58	
3	44.11	76.26	
4	86.40	68.17	
5	70.10	86.75	
Average	**49.88**	**78.74**	

Table 4.
The results of hemispherical photography photo analysis using Image-J software.

No	Station	Canopy cover (%)	Category
1	MDRT01	81.64 ± 3.35	Good
2	MDRT02	64.31 ± 20.41	Moderate

Table 5.
Canopy cover percentage and damage categories.

criteria and guidelines for determining mangrove damage. Based on the percentage canopy cover, the mangrove ecosystem in Tajungan Village, Kamal District, Bangkalan Regency is in the moderate-to-good category. The amount of mangrove ecosystem canopy cover at each station and the average cover, along with the standard deviation, can be seen in **Figure 6**.

As one of the suitability criteria for mangrove ecotourism in the research area, mangrove cover is in the appropriate category at the MDRT02 station and very suitable at the MDRT01 station. The category corresponds to the 50–75% coverage range and is very suitable in the >75% range [10]. The higher the percentage of mangrove canopy cover, the higher the level of health and suitability for ecotourism.

The tree canopy functions like an umbrella, dividing and breaking the penetration of sunlight and rain. Dominant mangrove species affect the percentage of mangrove canopy cover. Areas dominated by *Rhizophora* sp. with broad leaf morphology had a more significant percentage of canopy cover than areas dominated by mangrove species with small leaves. Stem diameter, density, and tree height also determine the level of mangrove canopy cover [18]. Global trends show that rainfall, temperature, cyclone frequency, and other geophysical factors that affect the maximum mangrove canopy height by 74% on a local and regional scale [19]. In addition, environmental damage due to sea waves, sunlight levels, and predation can affect the formation of

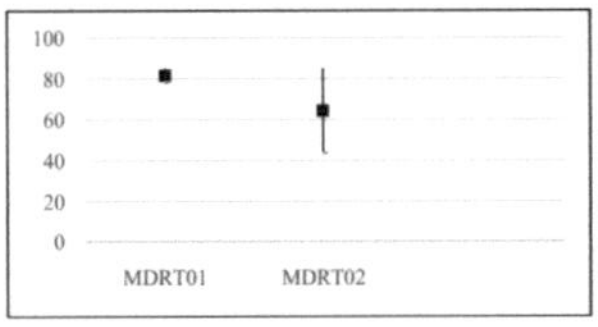

Figure 6.
The average percentage of cover and standard deviation at each observation station.

mangrove canopy cover [16]. The large tree diameter with high-density supports canopy cover, influencing the mangrove cover percentage. Mangrove canopy cover can show the natural level of mangrove ecosystems and detect anthropogenic threats [17]. In addition, the tree density value supports the mangrove cover's relatively good condition [20]. Based on this, tree categories' density and environmental characteristics' suitability generally affect the percentage of mangrove canopy cover [20, 21]. The results of this study can be used as a guide in developing educational ecotourism planting mangrove seedlings.

4. Conclusions

1. Analysis of canopy cover in the mangrove ecosystem in Tajungan Village, Kamal District, Bangkalan Regency, using the hemispherical photography method, shows that the highest percentage of canopy cover is at the MDRT01 station (81.64%) with the good category and the MDRT02 station (64.31%) with the moderate category.

2. The suitability of tourism for the percentage of cover category is in the appropriate category at the MDRT02 station and very suitable at the MDRT01 station.

Author details

Maulinna Kusumo Wardhani
Trunojoyo University of Madura, Bangkalan, Indonesia

*Address all correspondence to: maulinna@gmail.com

References

[1] Kuenzer C, Bluemel A, Gebhardt S, Quoc TV, Dech S. Remote sensing of mangrove ecosystems: A review. Remote Sensing. 2011;**3**:878-928. DOI: 10.3390/rs3050878

[2] Giri C, Long J, Abbas S, Murali RM, Qamer FM, Pengra B, et al. Distribution and dynamics of mangrove forest of South Asia. Journal of Environmental Management. 2015;**148**:101-111. DOI: 10.1016/j.jenvman.2014.01.020

[3] Masood H, Afsar S, Zamir U, Bin K, JH. Application of comparative remote sensing techniques for monitoring mangroves in Indus Delta, Sindh, Pakistan. Biological Forum—An International Journal. 2015;**7**(1):783-792

[4] Goldberg L, Lagomasino D, Thomas N, Fatoyinbo T. Global declines in human-driven mangrove loss. Global Change Biology. 2020;**26**:5844-5855. DOI: 10.1111/gcb.15275

[5] Hagger V, Worthington TA, Lovelock CE, Adame MF, Amano T, Brown BM, et al. Drivers of global mangrove loss and gain in social-ecological systems. Nature Communications. 2022;**13**(1):6373. DOI: 10.1038/s41467-022-33962-x

[6] Khakim N, Musthofa A, Wicaksono A, Lazuardi W, Pratama DND, Marfai MA. Adaptation of mangrove ecotourism management to coastal environment changes in the special region of Yogyakarta. Journal of Environmental Management and Tourism. 2021;**12**(3):754-765. DOI: 10.14505//jemt.12.3(51).14

[7] Wardhani MK. Mangrove ecotourism opportunity spectrum at south coast of Bangkalan Province of East Java. In: Proceedings of the 1st International Conference on Tropical Biology;4-5 Oktober 2013; SEAMEO BIOTROP Bogor. 2013. pp. 100-109

[8] Ziegler J, Araujo G, Labaja J, Snow S, King J, Ponzo A, et al. Can ecotourism change community attitudes towards conservation? Oryx. 2021;**55**(4):546-555. DOI: 10.1017/S0030605319000607

[9] Hakim L, Siswanto D, Nakagoshi N. Mangrove conservation in East Java: The ecotourism development perspectives. Journal of Tropical Life Science. 2017;**7**:277-285. DOI: 10.11594/jtls.07.03.14

[10] Bachmid KN, Schaduw JNW, Warouw V, Darwisito S, Kaligis EY, Wantasen A. Kajian Kesesuaian Lahan Ekowisata Mangrove Dimensi Ekologi (Kasus Pada Pulau Bunaken Bagian Timur, Kelurahan Alung Banua, Kecamatan Bunaken Kepulauan, Kota Manado). Jurnal Pesisir dan Laut Tropis. 2019;**7**(3):129-141

[11] Wardhani MK, Rosyid DM, Armono HD. Land use change of mangrove forest for eco-tourism in the south coastal, Bangkalan, East Java-Indonesia. GEOMATE Journal. 2022; **23**(98):136-146. Available from: https://geomatejournal.com/geomate/article/view/1221

[12] Decree of the Minister of Environment No. 201 of 2004

[13] Jenning SB, Brown ND, Sheil D. Assessing forest canopies and understorey illumination: Canopy closure, canopy cover, and other measures. Forestry. 1999;**72**(1):59-74. DOI: 10.1093/forestry/72.1.59

[14] Korhonen I, Korhonen KT, Rautiainen M, Stenberg P. Estimation of

forest canopy cover: A comparison of field measurement techniques. Silva Fennica. 2016;**40**(4):577-588. DOI: 10.14214/sf.315

[15] Dharmawan IWE. Hemispherical Photography. Analisis Persentase Tutupan Kanopi Komunitas Mangrove. CV. Makassar: Nas Media Pustaka; 2020

[16] Tomlinson PB. The Botany of Mangrove. New York: Cambridge; 1994

[17] Dharmawan IWE, Pramudji D. Panduan Monitoring Kesehatan Ekosistem Mangrove. Jakarta: COREMAP-CTI, Pusat Penelitian Oseanografi, LIPI; 2014

[18] Kuncahyo I, Pribadi R, Pratikto I. Komposisi dan Tutupan Kanopi Vegetasi Mangrove di Perairan Bakauheni Kabupaten Lampung Selatan. Journal of Marine Research. 2020;**9**(4):444-452. DOI: 10.14710/jmr.v9i4.27915

[19] Simard M, Fatoyinbo L, Smetanka C. Mangrove canopy height globally related to precipitation, temperature and cyclone frequency. Nature Geoscience. 2020;**12**:40-45. DOI: 10.1038/s41561-018-0279-1

[20] Nurdiansah D, Dharmawan IWK. Community Mangrove Alas Purwo National Park. Banyuwangi: National Parks Nature Center Purwo Banyuwangi; 2018

[21] Tinambunan SA, Pertami ND, Ernawati NN. Percentage of mangrove canopy cover and mollusks abundance in Benoa Bay mangrove ecosystem. Advances in Tropical Biodiversity and Environmental Sciences. 2021;**5**(3):97-102. DOI: 10.24843/atbes.v05.i03.p05

Chapter 6

Mangroves and Ecosystem-Based Coastal Protection in the Mekong River Delta, Vietnam

Klaus Schmitt and Thorsten Albers

Abstract

Development and the unsustainable use of natural resources in the coastal zone of the Mekong Delta, Vietnam, as well as erosion are threatening the protection function of the mangrove forests which protect the land behind the dyke from flooding and storms and provide co-benefits and livelihood for people in the coastal zone. These threats will be exacerbated by the impacts of climate change. Tidal flats and mangrove forests are an energy conversion system that provides ecosystem-based coastal protection. In sites where the mangrove belt has been destroyed and the tidal flat eroded, restoration of the tidal flats is a precondition for mangrove rehabilitation. Permeable bamboo fences, arranged in a T-shape, are effective for reducing erosion, stimulating sedimentation and thereby restoring tidal flats and re-creating conditions for mangrove regeneration. This cost-effective approach is only feasible within specific boundary conditions. Mangroves need to be protected from future anthropogenic destruction. This can best be achieved though co-management with the local people.

Keywords: ecosystem-based adaptation, coastal protection, erosion protection, mangrove regeneration, co-management

1. Introduction

The coastal zone of the Mekong Delta in Vietnam is facing cumulative challenges including the unsustainable use of natural resources, pollution, development, population growth and increased consumption. These challenges are exacerbated by the impacts of climate change—increased intensity of storms, flooding and sea level rise—resulting in erosion of the muddy coastlines [1–7].

As a result, 15.5% of the population of Vietnam is exposed to high coastal flood risk [8]. The traditional response to erosion and flooding is coastal protection through dykes, revetments and seawalls. This is very expensive, does not work on soft soils of mud coasts [9] and the possibility of increasing the dyke height is also limited due to the load bearing capacity of the muddy soil. The construction of concrete coastal protection elements may lead to maladaptations, or path dependencies [10, 11]. Wave attenuation by mangroves is an effective use of ecosystem services that protects dykes from erosion and the land behind the dykes from flooding, storms and sea levels rise.

In sites where erosion has eroded the foreshore and destroyed the mangrove forest in front of the dyke, tidal flat management is required to restore the eroded tidal flat. This will create the pre-conditions for regeneration or rehabilitation of mangrove forests.

This can be achieved through appropriate and site-specific approaches to coastal protection. Over the last decade more and more literature has become available on this topic and authors use different terms to describe coastal protection systems that incorporate natural elements such as mangroves and tidal flats: area coastal protection [12], ecosystem-based coastal defence [13], ecological engineering [14, 15], building with nature [16, 17], engineering with nature [18], nature-based coastal defence structures [19]. All these solutions involve mangroves, which in 2011 [20] were described as important physical ecosystem engineers that can control sedimentation processes and coastal protection.

Coastal ecosystems provide cumulative benefits [21] and mangroves contribute to this by providing a wide range of ecosystem services [22, 23] which include shoreline stabilisation and protection of coastal areas from wave impacts and storms [24–29]. Using these ecosystem services can contribute to adaptation pathways that lessen cumulative pressures on coastal areas and livelihoods [30, 31]. Ecosystem-based (or area) coastal protection considers the whole area of the tidal flats and mangrove forests as an "energy conversion system" and is therefore a very effective ecosystem-based system for coastal protection. Seagrass beds and/or coral reefs become part of the area coastal protection system in sites where they grow.

Muddy tidal flats are an important stabilising element of the coastal protection system. They decrease the incoming wave energy and thereby protect the coast from flooding and erosion. The higher the tidal flat, the greater the wave dissipation capacity. This results in a considerable decrease in the wave load at the dyke. The wave reduction effect is even bigger when mangrove forests grow on the tidal flats. The resulting decrease in wave height and length leads to a shortened wave run-up which decreases the dyke height needed and thereby lowers construction costs [32–35].

Vietnam is one of the six Southeast Asian nations where up to 80% of the 62% of global human-driven mangrove losses between 2000 and 2016 occurred [6]. The main anthropogenic drivers in Southeast Asia are conversion of mangrove forests to aquaculture and agriculture followed by logging. Once degraded or destroyed, the process of natural erosion is exacerbated [36]. Rates of mangrove loss have been slowing in recent years, suggesting that the importance of mangroves is becoming more widely recognised and that better management practices are being put in place [37, 38]. Nevertheless, good management practices are still often neglected in favour of mangrove planting to offset historic and ongoing mangrove loss. This can lead to malpractices in mangrove planting [39–42] and highlights the need for more effective mangrove conservation.

Conservation, in the sense of protection and management, of existing ecosystems and of managed land is more effective than rehabilitation.[1] Protection and management contributes 80% of the potential for cost effective climate mitigation from

[1] The terms rehabilitation and restoration are often used synonymously, but they have distinct meanings and are used in this chapter accordingly. Rehabilitation means "to make suitable again" while restoration to rebuild, to re-establish. In an ecological context, rehabilitation refers to "return ... degraded mangrove land to a fully functional mangrove ecosystem regardless of the original state of the degraded land", or in other words to convert a degraded system to a more stable condition ([43], p. 47).

Figure 1.
Eroded tidal flat. The dyke protection with concrete and Melaleuca fences failed to stop the erosion (Nopol, Soc Trang Province, Mekong Delta, Viet Nam, photo K. Schmitt 2010).

Nature-based Solutions[2] on land [45]. The most effective pathway therefore is to maintain the health of existing mangrove forests and reduce the rate of mangrove destruction or degradation. This can best be achieved though the participatory involvement of local people and co-management or shared governance [46–50]. Large-scale planting of mangroves in contrast, may increase the mangrove area in the short-term, but the long-term effectiveness is limited, and involves the risk of being used as an offset for the continued destruction of existing functional and diverse mature forests. Mangrove planting, using the wrong species in the wrong sites, may also result in collateral damage to existing or adjacent habitats, biodiversity trade-off and negative impacts on the local population [41, 51–53].

Over the last at least 75 million years [54], mangroves have developed unique characteristics to cope with shoreline evolution which do not necessarily follow succession of other forest types [55, 56]. Mangrove foresters therefore need a sound understanding of mangrove ecology but also of coastal processes (waves, tides, currents and sediment transport), hydrology and morphodynamics (spatial and temporal), and use it for conservation, planting and management decisions [50, 51]. Mangroves are well-adapted to dynamic tropical coasts that are subject to destructive storms and generally recover quickly from both minor and major periodic disturbances through natural regeneration, without the need for planting [57–61]. In contrast, human interventions, such as dykes, dams and upstream hydropower developments, usually lead to permanent changes which may create conditions which are unsuitable for natural regeneration of mangroves.

Along muddy tropical coastlines and estuaries where severe erosion or human impact has destroyed the mangrove belt, restoration of tidal flats and their fine sediment balance is a precondition for mangrove regeneration or rehabilitation [12, 16, 35, 62–64].

---

2 Nature-based Solutions (NbS) are actions addressing key societal challenges through the protection, sustainable management and restoration of both natural and modified ecosystems, benefiting both biodiversity and human well-being [44].

This chapter will describe how eroded tidal flats (**Figure 1**) can be restored using bamboo T-fences as a cost-effective ecosystem-based solution which re-creates the site conditions suitable for mangrove regeneration or rehabilitation and, in addition, provides co-benefits, biodiversity conservation, and human wellbeing. The chapter also briefly explains how mangrove forests can be protected and sustainably managed and thereby reducing the risk of mangrove degradation or destruction in the future.

2. The T-fences

Systematic land-reclamation work using breakwaters has been carried out in the Wadden Sea in Germany, The Netherlands and Denmark since the eighteenth century [65, 66]. Restoration of tidal flats, with the aim of area coastal protection, using the same principles, namely T-shaped fences, was adapted to the situation in the Mekong Delta using local materials. The most effective design of breakwaters was tested and permeable, T-shaped bamboo fences filled with soft brushwood provided the best results [67]. In other areas of the Mekong Delta cost parallel Melaleuca fences were used [68] and a comprehensive overview of managing erosion of mangrove-mud coasts with permeable dams from 5 countries in Asia and South America is provided in [64].

In the Lower Mekong Delta, a total of 7500 m of permeable T-shaped bamboo fences were installed on the east coast in Soc Trang and Bac Lieu Provinces [34]. In addition, 925 m were installed in Ca Mau Province between 2015 and 2016 (**Figure 2**).

Before placing any structures in the sea, it is important to have a sound understanding of coastal processes, hydrology and morphodynamics. This, and monitoring

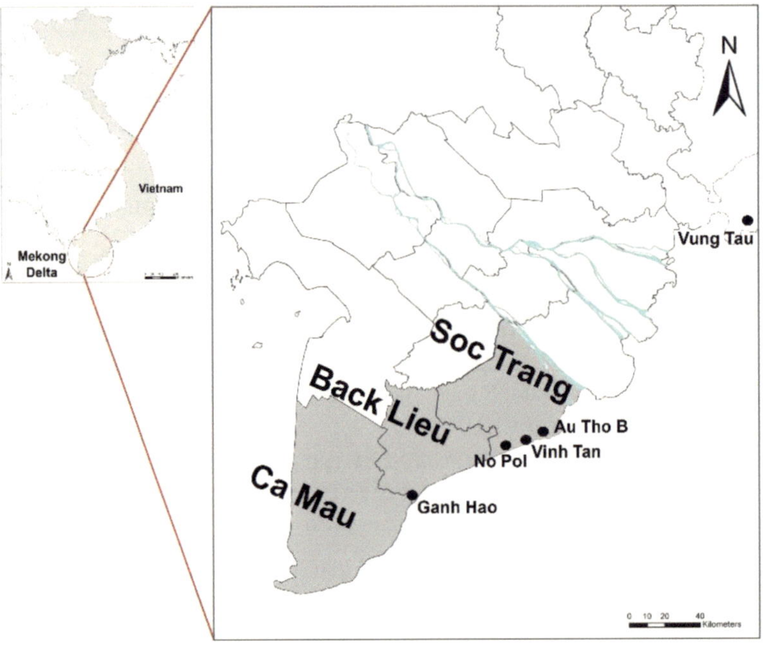

Figure 2.
Map of field sites in the Mekong Delta, Viet Nam; sluice gate 4 is about 1 km southwest of No Pol.

their impact on tidal flat restoration, will ensure that design specifications are appropriate for the site and that lee erosion can be minimised.

2.1 Numeric modelling

Numeric modelling of hydro- and sediment-dynamics provides the sound understanding and projecting of natural forces which are shaping the shoreline in order to plan the optimal placement as well as providing important boundary conditions for the design and construction of the T-fences. Information about the wave climate is essential when designing the bamboo fences. However, field measurements of waves cannot cover all weather conditions. Therefore, a numerical wave model SWAN (Simulating WAves Nearshore, www.swan.tudelft.nl) was setup, calibrated and verified to obtain the missing information using available data and data from field measurements from Vinh Tan. The numeric modelling was done in three steps. In a larger investigation area of approximately 250 km in north-south and 40 km in west-east-direction a wave model was set up from Vung Tau to Ganh Hao. The results were used as design parameters for the bamboo fences. The SWAN model was then coupled with the hydrodynamic model RMA-Kalypso (http://kalypso.wb.tu-harburg.de) to simulate currents and wave-induced currents. The results were used as input in the morphodynamic model GENESIS (Generalised Model for Simulating Shoreline Change) to simulate the shoreline changes [69] in the area of Vinh Tan based on the current and wave regimes. Structural measures such as conventional breakwaters but also the bamboo T-fences were integrated in the model and the resulting effects were simulated.

Boundary condition data on tides and wind from existing stations together with data on currents, waves, sediment concentrations and bathymetry recorded in the field were used to gain and improve the knowledge about hydrodynamic and morphodynamic processes [4, 12, 67, 70].

The modelling showed that recreating the former coastline by connecting existing headlands as shown in **Figure 3** will minimise lee erosion. The idealised shoreline is a relatively stable morphologic situation which often indicates the former shoreline. Closing the eroded gaps in the mangrove belt will create a "close to natural" situation without significant downdrift erosion.

2.2 Planning, design and construction of appropriate breakwaters

When designing the most effective structures to restore eroded tidal flats, their design, positioning and arrangement needs to be tested. This process started with an experimental design test and was afterwards tested and modified in the field.

Figure 3.
Placement of T-fences to minimise lee erosion (Bac Lieu Province, Viet Nam, photo Cong Ly and G.E. Wind 2013).

The wave dampening effects of conventional breakwaters (rubble mounds) and different designs using bamboo were tested in a wave flume: 2 rows of spaced bamboo poles, 4 rows of densely packed bamboo poles, and 2 rows of bamboo poles with brushwood in between (the latter is shown in **Figure 4**). Bamboo was selected due to its strength, local availability and costs [71]. The densely packed design was based on bamboo fences constructed along the Upper Gulf of Thailand (**Figure 5**) since 2005 [72].

The design with 2 rows of bamboo poles with brushwood in between provided the best results (**Figure 6**) and was therefore tested in the field. Different installation techniques were used to find the most efficient construction method. This included the application of a manual head ram, pressure using the weight of several people and pressure combined with vibrations, at a later stage an excavator on a pontoon was used to push the poles with the excavator shovel into the mud. Tensile tests were carried out with single and groups of poles until failure to verify the material parameters used in the theoretical design of the bamboo fences. The optimum diameter was derived from the design approach and the tensile tests. Also the calculated depths of embedment could be verified. During the first field tests different tying materials (ropes, hemp rope, rattan and stainless-steel wire) and tying techniques were tested in order to find an optimised design and construction method [67].

Two designs were installed at the coast in Soc Trang in 2011, a double row of bamboo fences filled with soft and one filled with stiff brushwood bundles. Wave height measurements were carried out for about 6 months to quantify the wave transmission effect of the fences during various storm and tidal conditions. Pressure transducers were installed 5 m from the fence on the sea- and landward side. The data were analysed and then summarised in significant wave heights of periods of 15 min [12].

The comparison of the results of the wave dampening effect of the physical model in the wave flume and field measurements are summarised in **Figure 6**. It shows the wave transmission coefficient k_T in relation to a quotient of the freeboard R_C and the initial significant wave height H_S. The solid lines represent the best-fit through the measured values. The black triangles are the results of the physical modelling while the red squares and blue Xs are the results of the field measurements. Flexible bundles lead to smaller wave transmission coefficients than stiff bundles, and thus

Figure 4.
Physical modelling of wave transmission through bamboo fence, scale 1:20 (photo T. Albers 2010).

Figure 5.
Bamboo fences in Khok Kha, Samut Sakhon Province, Thailand (photo K. Schmitt 2011).

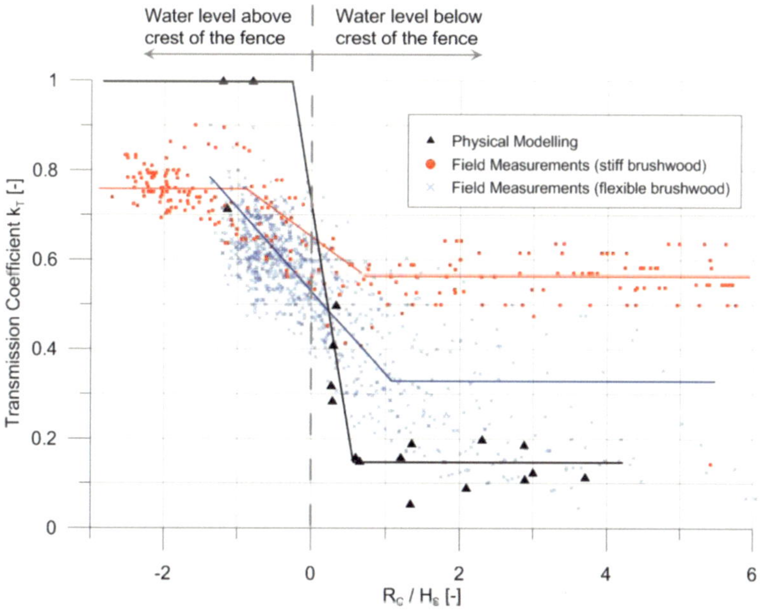

Figure 6.
Three different scenarios of wave transmission coefficients of bamboo fences under various hydrological conditions (modified from [70]).

have a larger wave dampening effect. They can reach up to an 80% reduction of the initial wave height. This was also confirmed by [73] who concluded that fence porosity drastically affects attenuation of both high- and low-frequency waves [74] applied the numerical model SWASH to simulate the wave transmission of bamboo fences.

Although the model showed transmission coefficients that were up to 30% higher than in the field study, i.e., lower wave reduction than measured, there are matching trends between the simulation results and the field measurements due to different input parameters.

The arrangement of the permeable bamboo fences consists of a long-shore and a cross-shore part. The long-shore parts dampen the incoming wave energy and the cross-shore parts decrease the long-shore currents as can be seen in **Figure 7**.

Flow and sediment transport patterns through the permeable fences and the gaps improve sediment input and accelerate the sediment consolidation process. The long-shore fences break the waves and the cross-shore parts catch sediments suspended in long-shore currents. The gaps in the long-shore fences increase sediment input into the fields created by the fences during flood tide. During ebb tide, drainage is accelerated through the gaps, and this increases the speed of the soil consolidation process in the fields (**Figure 8**).

2.3 Fence design, boundary conditions and monitoring

The results of the field measurements and the numeric modelling, and the analysis of sediment accretion monitoring and natural regeneration of mangroves, as well as maintenance data from construction sites of bamboo T-fences were used to define the design and boundary conditions of the bamboo fences.

The fences consist of two rows of vertical bamboo poles with a mean diameter of 8 cm and brushwood bundles in the gap. The distance between the two rows is 0.40 m for cross-shore sections and 0.50 m for the long-shore sections. The distance between the vertical poles is about 0.30 m. A double row of horizontal poles is connected to the vertical poles on each side. The brushwood bundles consist of small, soft bamboo branches. Stainless steel wire is used to tie the joints. A double layer of *Nypa* palm leaves was installed to reduce scouring at the bottom of the fences (**Figure 9**).

Figure 7.
Wave dampening effect of bamboo T-fences, Ca Mau Province, Viet Nam (photo R. Sorgenfrei 2016).

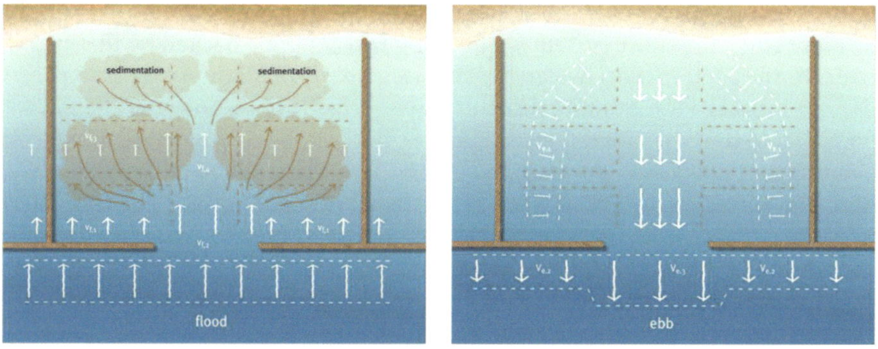

Figure 8.
Flow patterns and sediment transport in the fields protected by the fences (V_f = current velocity during flood tide, V_e = current velocity during ebb tide) (from [70]).

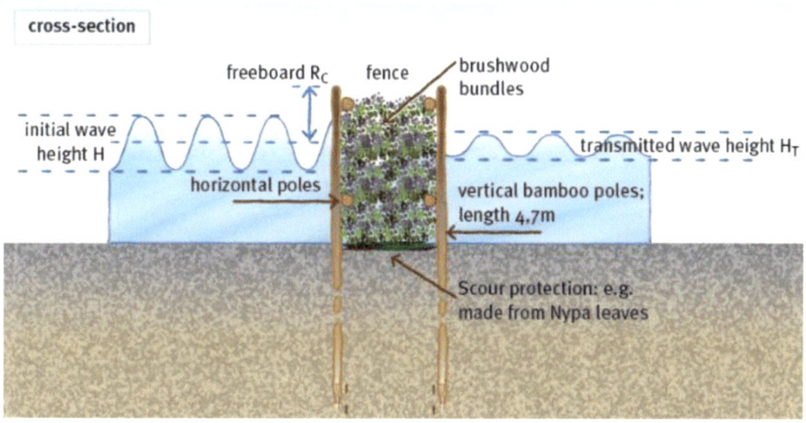

Figure 9.
Design of the permeable bamboo fences and resulting wave transmission (from [70]).

However, scouring cannot be completely avoided and thus the depth of embedment of the vertical poles was chosen to be large enough so that local scouring does not affect the stability of the fences. In the case of the muddy coast in Soc Trang, this was 3.4 m with about 0.8 m embedded in mud and about 2.6 m in sand.

The breaking force of the bamboo was estimated based on a literature review and verified by the tensile tests. The calculation of the loads on the front row of the bamboo fence resulting from current forces and acceleration forces of the tidal current as well as waves was done based on the superposition method by Morison, O'Brian, Johnson and Schaaf [75]. The rear row of the bamboo fence is loaded by the horizontal current- and tide-induced forces transmitted by the brushwood wall. The calculation of the resulting loads was done with the Coastal Engineering Design and Analysis System (CEDAS—https://www.veritechinc.com/products/cedas) based on the approaches of Miche-Rundgren and Sainflou [76] also considering slamming forces of breaking waves. Abnormal forces can result from the impact of floating items like flotsam or vessels. To address this, an impact of a 300 kg item was taken into account. Additionally a man weight of 1 kN as a vertical load was assumed for each bamboo pile.

The bamboo poles transfer horizontal loads to the ground by an elastic clamping of the pole. Thus, the static system is a bending resistant pile backed by the surrounding soil. For the geotechnical design the subgrade reaction method was used [77]. It is inferred that the horizontal pressure between the bamboo pole and the soil is proportional to the horizontal displacement of the pole. The proportionality factor k_s (bedding modulus) can vary with the depth. In this case the parable of Titze was applied, that offers a good description of the distribution of k_s [78]. The characteristics of the sand layer were used for the geotechnical design. The embedment depth is thus the depth in the sand layer. The mud layer is considered as a buffer layer that can grow and shrink due to external factors such as increase or decrease of incoming wave energy and does not have load-bearing attributes.

Disintegrating bamboo structures in the Upper Gulf of Thailand release floating debris which damages mangrove tree stems [72]. This problem has not been observed in the Mekong Delta where much less bamboo is used for the breakwaters then in Thailand (see **Figure 5**). Furthermore, the embedment depths is more than 2 times the above ground fence height, the poles are connected with stainless steel wire and monitoring and life-cycle-management ensures proper functioning of the infrastructure component. This minimises the risk of mangrove damage through floating debris. In addition, the effect of floating items with an impact of a 300 kg was considered in the fence design.

The following boundary conditions must be fulfilled to ensure that the fences, as described above, can be applied successfully:

- Muddy environment; medium grain size diameter of top layer of the mud $d_{50} < 0.03$ mm
- Significant wave height $H_s < 0.90$ m
- Mean wave period $T_m < 8$ s
- Small gradient of the tidal flat <1:1000
- Hight of the fence <1.40 m (while the crest height of the fence is equivalent to the mean high-water level during spring tide)

These five boundary conditions are summarised in **Figure 10**. The x-axis shows 2 parameters, namely significant wave height H_s and mean wave period T_m.

Only if all parameters measured are within the blue rectangle with rounded corners is the application of bamboo T-fences feasible. The colour gradient in the rectangle indicates that there is no clear boundary of applicability. If the limiting criteria are exceeded to some extent, adaptations, such as strengthening with concrete poles, must be considered. If the limiting criteria are greatly exceeded, an application of T-fences is not feasible.

There are additional limiting factors which should be considered. The thickness of the top mud layer indicates the amount of sediments in the system to restore the eroded tidal flats. In the Mekong Delta, > 0.50 m of mud layer has shown to be sufficient at providing enough sediment to restore sever erosion (**Figure 1**, the picture in **Figures 11** and **12**).

Further, it must be considered that bamboo attracts shipworms (wood-burrowing bivalves with wormlike bodies, *Teredo* sp. and *Bankia* sp.). In sites with steep shoreline

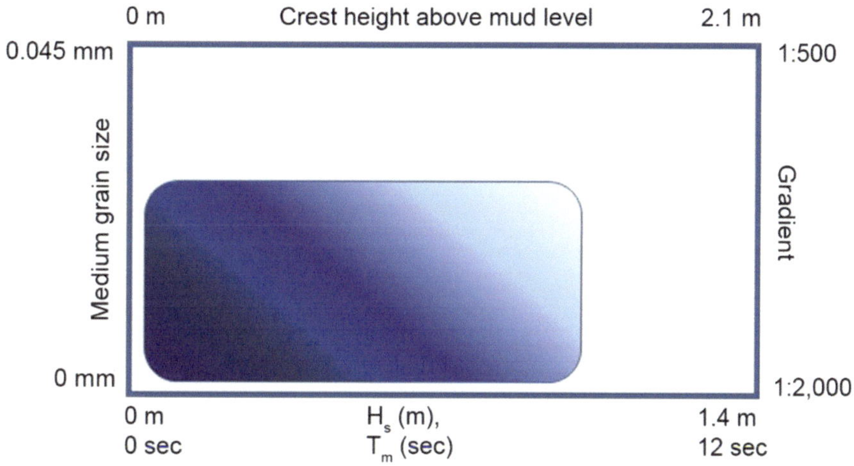

Figure 10.
Five key boundary conditions within which application of bamboo T-fences is feasible (modified from https://panorama.solutions/en/solution/ecosystem-based-coastal-protection-through-floodplain-restoration).

Figure 11.
Natural regeneration of Avicennia on restored tidal flats at Sluice Gate 4 in Soc Trang Province from the construction of the T-fences in October 2012 until January 2015 (photos: GIZ Soc Trang, R. Sorgenfrei).

gradients and long submergence periods, shipworms affected or even destroyed the T-fence structure after a few months. The risk of shipworm attack can be minimised by building the fences within the appropriate boundary conditions.

The duration of submergence and exposure to waves also affect the effort required for maintenance. Long submersion weakens the construction material and larger wave forces influence the stability of the connections. The longer the duration of

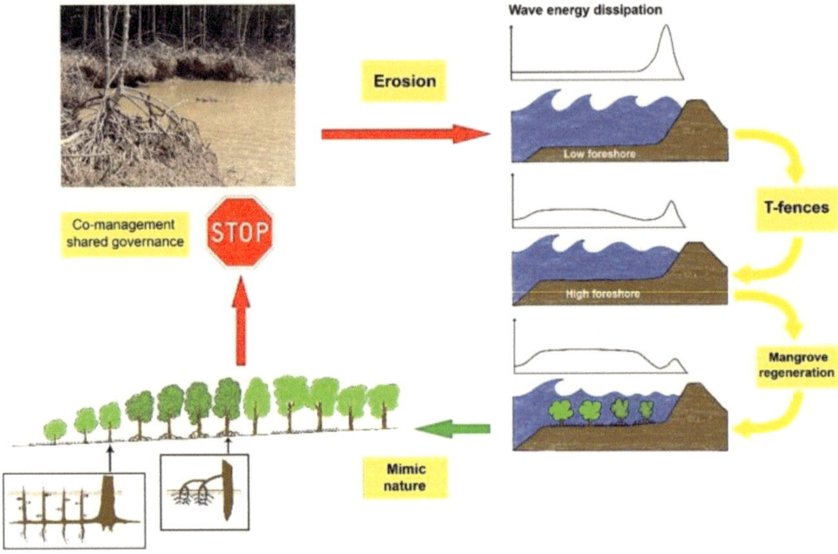

Figure 12.
The steps from eroded foreshore through flood plain restoration to mangrove regeneration/rehabilitation. Effective protection of the mangroves can prevent re-occurrence of erosion due to degradation or destruction of the mangroves (from [34]).

submergence and the higher the degree of exposure to waves, the larger is the effort required for maintenance. Of course, both input parameters correlate, since wave heights can be larger in deeper water.

T-fence monitoring and maintenance ensures proper functioning of the infrastructure component. During the first year after construction visual inspections should be carried out at monthly intervals and maintenance should be carried out where necessary. After that, visual inspections and maintenance should be carried out as a minimum after every storm season. Seasonal GPS (Global Positioning System) surveys of the shoreline at low tide can provide information if the T-fences have impacts on the shape of the nearby coastline.

2.4 Effects of T-fences

The reduction in wave height and thus in orbital velocities under waves and the flow and sediment transport in the fields created by T-fences leads to accelerated sedimentation rates [4, 70]. The reduction of wave action on the landward side of the fences also accelerates the consolidation of the mud and thus increases the stability of the sediments against erosion. The resulting restoration of the tidal flats creates the precondition for mangrove regeneration (**Figure 11**).

The 4 fixed-photo pictures in **Figure 11** were taken between 2012 and 2015. In November 2012 the coast parallel elements of the T-fences and the gap are still visible. In the foreground gabions are visible, placed at the front of the dyke to protect it from erosion and overtopping. In February 2013 the beginning of the sedimentation can clearly be seen on the left side of the picture. In November 2013 consolidation of sediments has started from the edge towards the gaps in the T-fences. This is indicated by the change in mud colour which is darker on the right were natural regeneration of *Avicennia* is already occurring. The photo taken in January 2015 shows the growth of mangroves, that are not disturbed by wave action (due to the high/restored tidal flat) and that are protected from destructive human impacts.

2.5 Costs and benefits

The costs for the construction of bamboo T-fences were about US$ 50–60 per meter in 2008, the costs per meter for a 3.5 m high concrete dyke were US$ 2270 [79], based on an average exchange rate in 2008 of 16,300 Vietnam Dong per US$.

The lifespan of bamboo fences (5–7 years, pers. comm. Worapol Douglomchan 2011, Khok Kha, Samut Sakhon Province, Thailand) is sufficient for the restoration of tidal flats at coasts with adequate supply of fine-grained sediment. If sediment- and morphodynamics change over time, bamboo T-fences—in contrast to concrete construction elements of coastal protection—can easily be adjusted.

A comprehensive review of economic values of mangrove ecosystem services is provided by [21]. In northern Vietnam, for example, an initial investment of USD 1.1 million in mangrove planting saved an estimated USD 7.3 million a year in sea dyke maintenance [80]. A study from Soc Trang compared the values of mangrove planting with a dyke upgrade based on saved wealth and saved health[3] [81]. The saved wealth index per USD invested for mangroves is about 19 times higher than for the dyke upgrade. In addition, mangroves are able to provide health benefits of 243 Disability-Adjusted Life Years in 20 years whereas the dyke upgrade does not deliver any positive health impacts [82].

3. Mangrove management

After successful restoration of sites suitable for mangrove growth, natural regeneration of mangroves will occur if environmental conditions are below key biophysical thresholds [12, 83]. If rates of natural regeneration are insufficient, supplementary planting of mangroves may be necessary. In such cases, appropriate species need to be planted at the right sites and at the correct time [12, 41, 84]. It is, however, essential to address the underlying factors leading to mangrove deforestation and degradation. This can best be achieved through effective protection and management of mangroves otherwise the cycle of anthropogenic degradation/destruction and expensive restoration will continue uninterrupted (**Figure 12**). Involving local people through co-management has shown to achieve this in an effective way which, in addition, provides co-benefits for the local population [46–49].

Mangrove co-management is based on participatory negotiation, joint decision-making, a degree of power-sharing, and a fair distribution of benefits among all stakeholders. It empowers local people to negotiate with local authorities and take over the management of mangroves. A partnership agreement between the resource users and local authorities will give the user group the right to use natural resources sustainably on a defined area of state-owned land (in the case of Vietnam Protection Forest) while being held responsible for the sustainable management and effective protection of those resources.

[3] Saved Wealth covers the monetary value of public infrastructure, private property and income loss; Saved Health covers avoided disease, disability and live loss, it is a concept to quantify the burden of disability and death, expressed as the number of years lost due to disability and early death.

In Au Tho B village in Soc Trang Province, mangrove co-management resulted in enhanced biodiversity, improved coastal protection and enhanced livelihoods through more income from fisheries as well as better collaboration between local people and local authorities [49]. The mangrove area under co-management in front of the village increased between 2008 and 2022 from about 70 to almost 280 ha without any planting.

4. Summary

Coastal areas are complex and dynamic ecosystems that face cumulative challenges and uncertainties due to human impacts and climate change. To address the uncertainties, complexity and adaptive capacity, a number of adaptation strategies should be used. These should contain different site specific and appropriate solutions to coastal protection and mangrove rehabilitation to avoid maladaptation, path dependencies and ultimately a reduction in adaptive capacity [31, 85, 86].

A diverse strategy which does not rely on concrete structures and which combines appropriate site-specific elements can respond in a flexible way to future scenarios about flow regimes and sediment patterns. The dynamic coastline of the Mekong Delta, for example, is largely influenced by sediment transport from the Mekong River which is predicted to diminish by 50% in 2050–2060 mainly due to hydropower development in the catchment area [87]. The need for a coastal defence strategy which is viable over time has also been identified as the solution for the dynamic mud-bank mangrove system along the coast of Guyana [88].

Knowledge of the main drivers of coastline changes and the way they influence the coastline and mangrove cover and of historical processes and coastal dynamics is also important for the development of adaptation strategies [5, 36, 89].

Fore shore management, including the stimulation of sedimentation using bamboo T-fences, is a cost-effective and sustainable approach, which does not cause any major interference with natural coastal morphodynamics if the placement of the T-fences more or less recreates the original coastline. The application therefore requires measurements of currents, waves, sediment concentrations and bathymetry as well as a sound understanding of mangrove ecology and coastal dynamics.

The wave transmission effect of bamboo T-fences is sufficient to significantly reduce wave heights and stimulate sedimentation on the landward side. The construction is cost-efficient and often more feasible than massive concrete structures on soft soil.

However, the application of T-fences has clear limits. It is only feasible within specific boundary conditions and T-fences must be sustained through a sound life-cycle-management including a maintenance strategy. If the site exceeds the amount of exposure to waves and duration of submergence, the effort for maintenance increases a lot and ultimately the use of T-fences becomes impractical. The applicability, design and layout of the T-fences, therefore, must be checked for every site and modified if required. For sites which exceed the limiting criteria to a large extent alternative solutions must be put in place.

It is essential that the mangroves are protected from human impacts once natural regeneration has occurred or mangroves have been planted otherwise the cycle of anthropogenic degradation/destruction and expensive restoration will continue. This can best be achieved by involving local people in effective protection and management of mangroves through co-management. Mangrove conservation can also supports the process of natural regeneration without the need for planting.

Ecosystem-based coastal protection using mangroves delivers a wide range of benefits. Mangrove forests provide co-benefits and livelihood for people living in the coastal zone. They contribute to protection from erosion, flooding, storms and rising sea levels. Furthermore, mangroves sequester greenhouse gases, protect biodiversity, provide a more economical solution to address coastal threats and can adapt to changing conditions.

Author details

Klaus Schmitt[1*] and Thorsten Albers[2]

1 Deutsche Gesellschaft für Internationale Zusammenarbeit (GIZ) GmbH, Manila, Philippines

2 Ostfalia—University of Applied Sciences, Suderburg, Germany

*Address all correspondence to: klaus.schmitt@giz.de

References

[1] Lee SY, Dunn RJK, Young RA, Connolly RM, Dale PER, Dehayr R, et al. Impact of urbanization on coastal wetland structure and function. Austral Ecology. 2006;**31**(2):149-163. DOI: 10.1111/j.1442-9993.2006.01581.x

[2] Carew-Reid J. Rapid Assessment of the Extent and Impact of Sea Level Rise in Viet Nam. Climate Change Discussion Paper 1. Brisbane: ICEM International Centre for Environmental Management; 2007. p. 74

[3] Schmidt-Thome P, Nguyen TH, Pham TL, Jarva J, Nuottimäki K. Climate change in Vietnam. In: Climate Change Adaptation Measures in Vietnam. SpringerBriefs in Earth Sciences. Cham: Springer; 2014. pp. 7-15. DOI: 10.1007/978-3-319-12346-2

[4] Schmitt K, Albers T. Area coastal protection and the use of bamboo breakwaters in the Mekong Delta. In: Nguyen DT, Takagi H, Esteban M, editors. Coastal Disasters and Climate Change in Vietnam: Engineering and Planning Perspectives. Elsevier Insights; 2014. pp. 107-132. DOI: 10.1016/B978-0-12-800007-6.00005-8

[5] Joffre OM, Schmitt K. A brief history of mangrove distribution and coastline development in Soc Trang Province, Vietnam, to address coastal management strategies. Advances in Global Change Research. 2019;**64**:67-85. DOI: 10.1007/978-3-319-90400-9_5

[6] Goldberg L, Lagomasino D, Thomas N, Fatoyinbo T. Global declines in human-driven mangrove loss. Global Change Biology. 2020;**26**(10):5844-5855. DOI: 10.1111/gcb.15275

[7] IPCC. Climate Change 2022: Impacts, Adaptation and Vulnerability. In: Pörtner H-O, Roberts DC, Tignor M, Poloczanska ES, Mintenbeck K, Alegría A, Craig M, Langsdorf S, Löschke S, Möller V, Okem A, Rama B, editors. Contribution of Working Group II to the Sixth Assessment Report of the Intergovernmental Panel on Climate Change. Cambridge: Cambridge University Press; 2022. p. 3056. DOI: 10.1017/9781009325844

[8] Rentschler J, Salhab M, Jafino BA. Flood exposure and poverty in 188 countries. Nature Communications. 2022;**13**(1):1-11. DOI: 10.1038/s41467-022-30727-4

[9] Kempfert H-G, Raithel M. Experiences on dike foundations and land fills on very soft soils. In: Proc. Intern. Workshop ISSMGE - Techn. Committee TC 36 "Foundation in Difficult Soft Soil Conditions". Mexico City; 2002. pp. 176-181

[10] IPCC. Climate change 2001: Impacts, adaptation, and vulnerability. In: McCarthy D, Canziani OF, Leary NA, Dokken DJ, White KS, editors. Contribution of Working Group II to the Third Assessment Report of the Intergovernmental Panel on Climate Change. Cambridge: Cambridge University Press; 2001. p. 1032. DOI: 10.1002/joc.775

[11] Barnett J, O'Neill S. Maladaptation. Global Environmental Change. 2010;**20**(2):211-213. DOI: 10.1016/j.gloenvcha.2009.11.00

[12] Schmitt K, Albers T, Pham TT, Dinh SC. Site-specific and integrated adaptation to climate change in the coastal mangrove zone of Soc Trang Province, Viet Nam. Journal of Coastal Conservation. 2013;**17**(3):545-558. DOI: 10.1007/s11852-013-0253-4

[13] Temmerman S, Meire P, Bouma TJ, Herman PMJ, Ysebaert T, De Vriend HJ. Ecosystem-based coastal defence in the face of global change. Nature. 2013;**504**(7478):79-83. DOI: 10.1038/nature12859

[14] Borsje BW, van Wesenbeeck BK, Dekker F, Paalvast P, Bouma TJ, van Katwijk MM, et al. How ecological engineering can serve in coastal protection. Ecological Engineering. 2011;**37**(2):113-122. DOI: 10.1016/j.ecoleng.2010.11.02

[15] Cheong S-M, Silliman B, Wong PP, van Wesenbeeck B, Kim C-K, Guannel G. Coastal adaptation with ecological engineering. Nature Climate Change. 2013;**3**(9):787-791. DOI: 10.1038/nclimate1854

[16] Winterwerp JC, Erftemeijer PLA, Suryadiputra N, Eijk P, Zhang L. Defining eco-morphodynamic requirements for rehabilitating eroding mangrove-mud coasts. Wetlands. 2013;**33**(3):515-526. DOI: 10.1007/s13157-013-0409-x

[17] de Vriend HJ, van Koningsveld M, Aarninkhof SGJ, de Vries MB, Baptist MJ. Sustainable hydraulic engineering through building with nature. Journal of Hydro-Environment Research. 2015;**9**(2):159-171. DOI: 10.1016/j.jher.2014.06.004

[18] Bridges TS, Bourne EM, King JK, Kuzmitski HK, Moynihan EB, Suedel BC. Engineering with Nature: An Atlas. ERDC/EL SR-18-8. Vicksburg, MS: U.S. Army Engineer Research and Development Center; 2018. p. 261. DOI: 10.21079/11681/2792

[19] Schoonees T, Gijón Mancheño A, Scheres B, Bouma TJ, Silva R, Schlurmann T, et al. Hard structures for coastal protection, towards greener designs. Estuaries and Coasts. 2019;**42**(7):1709-1729. DOI: 10.1007/s12237-019-00551-z

[20] Gutiérrez JL, Jones CG, Byers JE, Arkema KK, Berkenbusch K, Commito JA, et al. Physical ecosystem engineers and the functioning of estuaries and coasts. In: Wolanski E, McLusky DS, editors. Treatise on Estuarine and Coastal Science. Vol. 7. Waltham: Academic Press; 2011. pp. 53-81. DOI: 10.1016/B978-0-12-374711-2.00705-1

[21] Barbier EB, Hacker SD, Kennedy C, Koch EW, Stier AC, Silliman BR. The value of estuarine and coastal ecosystem services. Ecological Monographs. 2011;**81**(2):169-193. DOI: 10.1890/10-1510.1

[22] Millennium Ecosystem Assessment. Ecosystems and Human Well-Being: Wetlands and Water Synthesis. Washington DC, USA: World Resources Institute. 2005; 68. Available from: http://www.millenniumassessment.org/documents/document.358.aspx.pdf. [Accessed: January 18, 2023]

[23] Barbier EB. Valuing ecosystem services as productive inputs. Economy Policy. 2007;**22**(49):177-229. DOI: 10.1111/j.1468-0327.2007.00174.x/full

[24] Mazda Y, Magi M, Kogo M, Hong P. Mangroves as a coastal protection from waves in the Tong King Delta, Vietnam. Mangroves and Salt Marshes. 1997;**1**:127-135. DOI: 10.1023/A:1009928003700

[25] Alongi DM. Mangrove forests: Resilience, protection from tsunamis, and responses to global climate change. Estuarine, Coastal and Shelf Science. 2008;**76**:1-13. DOI: 10.1016/j.ecss.2007.08.02

[26] Zhang K, Liu H, Li Y, Xu H, Shen J, Rhome J, et al. The role of mangroves in attenuating storm surges. Estuarine, Coastal and Shelf Science. 2012;**102-103**:11-23. DOI: 0.1016/j.ecss.2012.02.021

[27] Lee SY, Primavera JH, Dahdouh-Guebas F, McKee K, Bosire JO, Cannicci S, et al. Ecological role and services of tropical mangrove ecosystems: A reassessment. Global Ecology and Biogeography. 2014;**23**:726-743. DOI: 10.1111/geb.12155

[28] Marois DE, Mitsch WJ. Coastal protection from tsunamis and cyclones provided by mangrove wetlands—A review. International Journal of Biodiversity Science, Ecosystem Services and Management. 2015;**11**(1):71-83. DOI: 10.1080/21513732.2014.997292

[29] Sandilyan S, Kathiresan K. Mangroves as bioshield: An undisputable fact. Ocean and Coastal Management. 2015;**103**:94-96. DOI: 10.1016/j.ocecoaman.2014.11.011

[30] Arkema KK, Guannel G, Verutes G, Wood SA, Guerry A, Ruckelshaus M, et al. Coastal habitats shield people and property from sea-level rise and storms. Nature Climate Change. 2013;**3**(10):913-918. DOI: 10.1038/nclimate1944

[31] Smith TF, Thomsen DC, Gould S, Schmitt K, Schlegel B. Cumulative pressures on sustainable livelihoods: Coastal adaptation in the Mekong Delta. Sustainability. 2013;**5**(1):228-241. DOI: 10.3390/su5010228

[32] Horstman EM, Dohmen-Janssen CM, Narra PMF, van den Berg NJF, Siemerink M, Hulscher SJMH. Wave attenuation in mangroves: A quantitative approach to field observations. Coastal Engineering. 2014;**94**:47-62. DOI: 10.1016/j.coastaleng.2014.08.005

[33] Spalding MD, McIvor AL, Beck MW, Koch EW, Möller I, Reed DJ, et al. Coastal ecosystems: A critical element of risk reduction. Conservation Letters. 2014;**7**(3):293-301. DOI: 10.1111/conl.12074

[34] Albers T, Schmitt K. Dyke design, floodplain restoration and mangrove co-management as parts of an area coastal protection strategy for the mud coasts of the Mekong Delta, Vietnam. Wetlands Ecology and Management. 2015;**23**(6):991-1004. DOI: 10.1007/s11273-015-9441-3

[35] Reed D, van Wesenbeeck B, Herman PMJ, Meselhe E. Tidal flat-wetland systems as flood defenses: Understanding biogeomorphic controls. Estuarine, Coastal and Shelf Science. 2018;**213**:269-282. DOI: 10.1016/j.ecss.2018.08.017

[36] Thomas N, Lucas R, Bunting P, Hardy A, Rosenqvist A, Simard M. Distribution and drivers of global mangrove forest change, 1996-2010. PLoS One. 2017;**12**(6):1-14. DOI: 10.1371/journal.pone.0179302

[37] Friess DA, Yando ES, Abuchahla GMO, Adams JB, Cannicci S, Canty SWJ, et al. Mangroves give cause for conservation optimism, for now. Current Biology. 2020;**30**(4):R153-R154. DOI: 10.1016/j.cub.2019.12.054

[38] Spalding M, Leal M. The state of the World's Mangroves. Global Mangrove Alliance. 2021;**2021**:79

[39] Turner R, Lewis R. Hydrologic restoration of coastal wetlands. Wetlands Ecology and Management. 1997;**4**(2):65-72

[40] Samson MS, Rollon RN. Growth performance of planted mangroves in the Philippines: Revisiting forest

management strategies. Ambio. 2008;**37**(4):234-240. DOI: 10.1579/0044-7447(2008)37[234:gpopmi]2.0.co;2

[41] Lee SY, Hamilton S, Barbier EB, Primavera J, Lewis RR. Better restoration policies are needed to conserve mangrove ecosystems. Nature Ecology and Evolution. 2019;**3**(6):870-872. DOI: 10.1038/s41559-019-0861-y

[42] Wodehouse DCJ, Rayment MB. Mangrove area and propagule number planting targets produce sub-optimal rehabilitation and afforestation outcomes. Estuarine, Coastal and Shelf Science. 2019;**222**:91-102. DOI: 10.1016/j.ecss.2019.04.003

[43] Field CD. Mangrove rehabilitation: Choice and necessity. Hydrobiologia. 1999;**413**:47-52. DOI: 10.1023/A:1003863415354

[44] Cohen-Shacham E, Walters G, Janzen C, Maginnis S. Nature-Based Solutions to Address Societal Challenges. Gland, Switzerland: International Union for Conservation of Nature; 2016. p. 79. DOI: 10.2305/IUCN.CH.2016.13.en

[45] Girardin CAJ, Jenkins S, Seddon N, Allen M, Lewis SL, Wheeler CE, et al. Nature-based solutions can help cool the planet - if we act now. Nature. 2021;**593**(7858):191-194. DOI: 10.1038/d41586-021-01241-2

[46] Borrini-Feyerabend G, Taghi Farvar M, Nguinguiri JC, Ndangang V. Co-Management of Natural Resources: Organising, Negotiating and Learning-by-Doing. Heidelberg, Kasparek Verlag; 2000. 95. Available from: https://www.researchgate.net/publication/327039413_Co-management_of_natural_resources_organising_negotiating_and_learning-by-doing. [Accessed: January 18, 2023]

[47] Borrini-Feyerabend G, Pimbert M, Farvar MT, Kothari A, Renard Y. Sharing Power: Learning by Doing in co-Management of Natural Resources throughout the World. Tehran: IEED and IUCN/CEESP/CMWG, Cenesta; 2004. p. 456. Available from: https://www.iied.org/sites/default/files/pdfs/migrate/G01089.pdf [Accessed: January 18, 2023]

[48] Erdman E, Merrill P, Mongdong M, Arsyad I, Harahap Z, Pangalila R, et al. Building Effective Co-Management Systems for Decentralized Protected Areas Management in Indonesia: Bunaken National Park Case Study. Jakarta, Indonesia: Natural Resources Management Program; 2004. p. 137

[49] Schmitt K. Mangrove planting, community participation and integrated management in Soc Trang Province, Viet Nam. In: Macinthosh DJ, Mahindapala R, Markopoulos M, editors. Sharing Lessons on Mangrove Restoration. Bangkok, Thailand: Mangroves for the Future and Gland. Switzerland: IUCN; 2012. pp. 205-225. Available from: http://mangrovesforthefuture.org/assets/Repository/Documents/Call-for-Action-and-Proceedings-from-2012-Colloquium-Mamallapuram-India.pdf. [Accessed: January 18, 2023]

[50] Hai NT, Dell B, Phuong VT, Harper RJ. Towards a more robust approach for the restoration of mangroves in Vietnam. Annals of Forest Science. 2020;**77**(1). DOI: 10.1007/s13595-020-0921-0

[51] Schmitt K, Duke NC. Mangrove management, assessment, and monitoring. In: Pancel L, Köhl M, editors. Tropical Forestry Handbook. 2nd ed. Vol. 2. Berlin, Heidelberg: Springer; 2016. pp. 1725-1759. DOI: 10.1007/978-3-642-54601-3_126

[52] Scheidel A, Work C. Forest plantations and climate change discourses: New powers of 'green' grabbing in Cambodia. Land Use Policy. 2018;**77**:9-18. DOI: 10.1016/j.landusepol.2018.04.057

[53] Lewis S, Wheeler CE, Mitchard ETA, Koch A. Regenerate natural forests to store carbon. Nature. 2019;**568**(7750):25-28. DOI: 10.1038/d41586-019-01026-8

[54] Gee CT. The mangrove palm Nypa in the geologic past of the new world. Wetlands Ecology and Management. 2001;**9**(3):181-203. DOI: 10.1023/A:1011148522181

[55] Duke NC, Ball MC, Ellison JC. Factors influencing biodiversity and distributional gradients in mangroves. Global Ecology and Biogeography Letters. 1998;**7**(1):27-47. DOI: 10.2307/2997695

[56] Alongi DM. Icarus revisited: Tropical forests, REDD+ and ecosystem dynamics. Carbon Management. 2013;**4**(5):469-472. DOI: 10.4155/cmt.13.4

[57] Jimenez J, Lugo A, Cintron G. Tree mortality in mangrove forests. Biotropica. 1985;**17**(3):177-185. DOI: 10.2307/2388214

[58] FAO. Mangrove Forest Management Guidelines. Rome: Food and Agriculture Organization of the United Nations; 1994. p. 321

[59] Alongi DM. The Energetics of Mangrove Forests. New York: Springer Science + Business Media B. V; 2009. p. 211

[60] Long J, Giri C, Primavera J, Trivedi M. Damage and recovery assessment of the Philippines' mangroves following Super Typhoon Haiyan. Marine Pollution Bulletin. 2016;**109**(2):734-743. DOI: 10.1016/j.marpolbul.2016.06.080

[61] Primavera JH, del Cruz M, Montilijao C, Consunji H, del Paz M, Rollon RN, et al. Preliminary assessment of post-Haiyan mangrove damage and short-term recovery in Eastern Samar, Central Philippines. Marine Pollution Bulletin. 2016;**109**(2):744-750. DOI: 10.1016/j.marpolbul.2016.05.050

[62] Winterwerp JC, Borst WG, de Vries MB. Pilot study on the erosion and rehabilitation of a mangrove mud coast. Journal of Coastal Research. 2005;**21**(2):223-230. DOI: 10.2112/03-832A.1

[63] Elliott M, Mander L, Mazik K, Simenstad C, Valesini F, Whitfield A, et al. Ecoengineering with ecohydrology: Successes and failures in estuarine restoration. Estuarine, Coastal and Shelf Science. 2016;**176**:12-35. DOI: 10.1016/j.ecss.2016.04.003

[64] Winterwerp JC, Albers T, Anthony EJ, Friess DA, Mancheño AG, Moseley K, et al. Managing erosion of mangrove-mud coasts with permeable dams – Lessons learned. Ecological Engineering. 2020;**158**:106078. DOI: 10.1016/j.ecoleng.2020.106078

[65] Probst B. Deichvorland bewirtschaftung im Wandel der Zeit. Die Küste. 1996;**58**:47-60

[66] Bakker J, Esselink P, Dijkema KS, van Duin WE, de Jong DJ. Restoration of salt marshes in the Netherlands. Hydrobiologia. 2002;**478**:29-51. DOI: 10.1007/978-94-017-1335-1_3

[67] Albers T, von Lieberman N. Current and Erosion Modelling Survey. Deutsche Gesellschaft für Internationale Zusammenarbeit (GIZ) GmbH, Management of Natural Resources in

the Coastal Zone of Soc Trang Province. Vietnam. 2011;**2011**:72. Available from: http://coastal-protection-mekongdelta.com/download/library/34.Current%20and%20Erosion%20Modelling%20Survey%20in%20Soc%20Trang_2011_EN.pdf. [Accessed: January 28, 2023]

[68] Van Cuong C, Brown S, To HH, Hockings M. Using Melaleuca fences as soft coastal engineering for mangrove restoration in Kien Giang, Vietnam. Ecological Engineering. 2015;**81**:256-265. DOI: 10.1016/j.ecoleng.2015.04.031

[69] Hanson H, Kraus NC. GENESIS: Generalized model for simulating shoreline change, report 1: Technical reference. Tech. Rep. CERC-89-19. U.S. Army Engineer Waterways Experiment Station, Coastal Engineering Research Center, Vicksbur. 1989; 185 p & 6 appendix. 1989. Available from: https://apps.dtic.mil/sti/citations/ADA217305. [Accessed: January 28, 2023]

[70] Albers T, San DC, Schmitt K. Shoreline management guidelines: Coastal protection in the lower Mekong Delta. In: Deutsche Gesellschaft für International Zusammenarbeit (GIZ) GmbH. Soc Trang, Vietnam; 2013. p. 124. Available from: http://coastal-protection-mekongdelta.com/download/library/40.Shoreline-Management-Guidelines_EN.pdf [Accessed: January 18, 2023]

[71] Halide H, Brinkman R, Ridd P. Designing bamboo wave attenuators for mangrove plantations. Indian Journal of Marine Sciences. 2004;**33**(3):220-225

[72] Pranchai A, Jenke M, Berger U. Well-intentioned, but poorly implemented: Debris from coastal bamboo fences triggered mangrove decline in Thailand. Marine Pollution Bulletin. 2019;**146**(July):900-907. DOI: 10.1016/j.marpolbul.2019.07.055

[73] Thieu Quang T, Mai TL. Monsoon wave transmission at bamboo fences protecting mangroves in the lower Mekong Delta. Applied Ocean Research. 2020;**101**:102259. DOI: 10.1016/j.apor.2020.102259

[74] Dao T, Stive MJF, Hofland B, Mai T. Wave damping due to wooden fences along mangrove coasts. Journal of Coastal Research. 2018;**34**(6):1317-1327. DOI: 10.2112/JCOASTRES-D-18-00015.1

[75] Sarpkaya T. Force on a circular cylinder in viscous oscillatory flow at low Keulegan-Carpenter numbers. Journal of Fluid Mechanics. 1986;**165**:61-71. DOI: 10.1017/S0022112086002999

[76] SPM. Shore protection manual. Coastal Engineering Research Center. DEPARTMENT OF THE ARMY Waterways Experiment Station, Corps of Engineers PC Box 631 Vicksburg, Mississippi 39180 r Vv H I 1984. 4th ed. Vicksburg, Mississippi, USA; 1984. p. 800. Available from: https://www.academia.edu/39978786/SHORE_PROTECTION_MANUAL_DEPARTMENT_OF_THE_ARMY_Waterways_Experiment_Station_Corps_of_Engineers_PC_Box_631_Vicksburg_Mississippi_39180_r_Vv_H_I_1984_Approved_For_Public_Release_Distribution_Unlimited_Prepared_for_DEPARTMENT_OF_THE_ARMY. [Accessed: January 18, 2023]

[77] Ziaie-Moayed R, Janbaz M. Effective parameters on modulus of subgrade reaction in clayey soils. Journal of Applied Sciences. 2009;**9**(2):4006-4012. DOI: 10.3923/jas.2009.4006.4012

[78] Terzaghi K, Peck RB, Mesri G. Soil Mechanics in Engineering Practice. 3rd ed. John Wiley & Sons. New York: Chichester, Brisbane, Toronto, Singapore; 1996. p. 512

[79] Hillen M. Safety Standards Project: Risk Analysis for New Sea Dike Design Guidelines in Vietnam. Delft University of Technology. Delft; 2008. p. 68. Available from: https://repository.tudelft.nl/islandora/object/uuid:59fd0624-9ad9-4eb7-ad9c-14164e904dfc/datastream/OBJ/download. [Accessed: January 18, 2023]

[80] Brown O, Crawford A, Hammill A. Natural disasters and resource rights, building resilience, rebuilding lives. International institute for. Sustainable Development. 2006;**2006**:28. Available from: https://www.iisd.org/system/files/publications/tas_natres_disasters.pdf [Accessed: January 18, 2023]

[81] Stadelmann M, Michaelowa A, Butzengeiger-Geyer S, Köhler M. Universal metrics to compare the effectiveness of climate change adaptation projects. In: Leal Filho W, editor. Handbook of Climate Change Adaptation. Berlin, Heidelberg: Springer; 2015. pp. 2143-2160. DOI: 10.1007/978-3-642-38670-1_128

[82] Köhler M, Michaelowa A. Saved health, saved wealth: An approach to quantifying the benefits of climate change adaptation. In: Deutsche Gesellschaft für Internationale Zusammenarbeit (GIZ) GmbH. Eschborn, Germany; 2013. p. 52. Available from: https://www.adaptationcommunity.net/?wpfb_dl=139. [Accessed: January 18, 2023]

[83] Friess DA, Rogers K, Lovelock CE, Krauss KW, Hamilton SE, Lee SY, et al. The state of the World's mangrove forests: Past, present, and future. Annual Review of Environment and Resources. 2019;**44**(1):89-115. DOI: 10.1146/annurev-environ-101718-033302

[84] Lovelock CE, Brown BM. Land tenure considerations are key to successful mangrove restoration. Nature Ecology and Evolution. 2019;**3**(8):1135. DOI: 10.1038/s41559-019-0942-y

[85] Munang RT, Thiaw I, Rivington M. Ecosystem management: Tomorrow's approach to enhancing food security under a changing climate. Sustainability. 2011;**3**:937-954. DOI: 10.3390/su3070937

[86] Magnan AK, Schipper ELF, Burkett M, Bharwani S, Burton I, Eriksen S, et al. Addressing the risk of maladaptation to climate change. Wiley Interdisciplinary Reviews: Climate Change. 2016;**7**(5):646-665. DOI: 10.1002/wcc.409

[87] Nguyen VM, Nguyen VD, Nguyen NH, Kummu M, Merz B, Apel H. Future sediment dynamics in the Mekong Delta floodplains: Impacts of hydropower development, climate change and sea level rise. Global Planet Change. 2015;**127**:22-33. DOI: 10.3390/su3070937

[88] Anthony EJ, Gratiot N. Coastal engineering and large-scale mangrove destruction in Guyana, South America: Averting an environmental catastrophe in the making. Ecological Engineering. 2012;**47**:268-273. DOI: 10.1016/j.ecoleng.2012.07.005

[89] Richards DR, Friess DA. Rates and drivers of mangrove deforestation in Southeast Asia, 2000-2012. Proceedings of the National Academy of Sciences of the United States of America. 2016;**113**(2):344-349. DOI: 10.1073/pnas.1510272113

Chapter 7

Perspective Chapter: Mangrove Conservation – An Ecotourism Approach

I. Ketut Ginantra

Abstract

The chapter we propose includes the following: The uniqueness of the mangrove ecosystem, the mangrove ecosystem is a unique ecotone, which connects the life of land and marine biota. Its diversity of plants and fauna is typical, namely, true mangrove plants, associated mangrove plants, crustaceans, mollusks, fish, reptiles and birds. The existence of mangrove flora and fauna is an attraction for ecotourism, scientific interpretation and educational materials for the community for mangrove conservation. Principles in conservation include preservation, protection and sustainable use. Ecotourism is the sustainable use of mangroves, which combines three main aspects, namely, ecology, economy and evaluating community opinion. Examples of the use of mangrove ecosystems are for ecotourism in Bali, Indonesia. The manuscript has a valuable contribution on the importance of the mangrove ecosystems in ecotourism.

Keywords: mangrove flora and fauna, conservation, attraction, ecotourism in Bali, mangrove tourism

1. Introduction

Mangroves are ecosystems that provide productive habitat and can support coastal fisheries including crabs, shrimp and fish, and have a high diversity of biota species. The fauna diversity includes arboreal, terrestrial, semi-aquatic fauna, mollusks, crustaceans, fish and other aquatic fauna. Mangroves are an excellent habitat to support growth and reproduction for the preservation of species in ecosystems. Mangroves are also a feeding ground, spawning ground and nursery ground for various associated marine biota [1, 2].

Ecotourism is a tourism activity that combines 3 main aspects, namely ecology, including the existence of the types that make up the mangrove ecosystem and also its conservation efforts. The second is Economic, the economic value generated from ecotourism activities in sustainable mangroves and part of the proceeds is returned to maintain the ecosystem and the third is the empowerment of the local (local) community as the initiator, manager and guide in the ecotourism business [3]. In principle, ecotourism is an activity in which the physical/chemical, biological/ecological and economic functions of mangrove forests continue to run well. Ecotourism in mangrove areas contains three main pillars, namely ecology, which is a vehicle for nature conservation, sustainable economy and empowerment of local (local) communities [3, 4].

The diversity of species and unique characteristics of mangrove plants and the diversity of fauna (birds, crustaceans, mollusks, fish) can be an attraction for ecotourism attractions, including plant species with unique roots, fruit shapes, and the adaptability of mangrove plants on muddy and able to live on land with high salt content. The benefits of plants for treatment, the benefits of mangrove plant species as mosquito repellents, or the benefits of plants for religious ceremonies in Bali can also be an attraction for ecotourism [5, 6].

The feasibility of mangrove forest ecosystems for ecotourism activities can be seen from the diversity of mangrove plant species (number of species), mangrove density (number of individuals/m2), mangrove thickness (mangrove diameter from coast to land), above-tree biota (insects, birds), biota in water (fish, crabs, mollusks), perceptions of local people and also the condition of mangrove forests [7].

The forms of ecotourism activities in Mangrove can be quite varied. Sports and recreational tourism activities, attractions can be in the form of kayaking, fishing, canoeing, camping. The facilities needed are kayaks, canoes, rafts, camping ground. Educational and research tourism, the attractions can be in the form of an introduction to mangrove vegetation, birth watching, an introduction to the characteristics of mangrove plants. The required facilities can be natural, canoe, raft, observation post/ecotower, resting point. Health tourism, attractions can be in the form of meditation, rehabilitation, therapy and the facilities needed are shelter, shade [3].

Mangrove forests with a diversity of unique flora and fauna are very attractive as a tourist attraction. Many mangrove areas have been developed as tourist attractions, including the Nusa Lembongan mangrove area for mangrove tours, the TAHURA Ngurah Rai mangrove area for ecotourism attractions, ecotourism areas in Kampoeng Kepiting mangrove forest, Pejarakan Buleleng village mangrove forest as an educational tourist attraction, mangrove forests in Perancak developed as an ecotourism attraction. The Segara Batu Lumbang mangrove forest is part of the Tahura Ngurah Rai mangrove forest area, which was developed by the Segara Guna Batu Lumbang Pemogan fishing group. The mangrove forest in Segara Batu Lumbang is also a tourist attraction based on the conservation of the diversity of mangrove flora and fauna. Mangrove tourism that has been developed is a mangrove tour with canoes, traditional boat "jukung", fishing tours, volunteer tourism.

Several other mangrove forest areas were also developed by local community groups including the Nusa Lembongan mangroves by the Sari Segara group, the mangrove forests on the coast of Pejarakan Buleleng by the Nature Conservation Forum Putri Menjangan and the Perancak mangrove forests by the Village-Owned Enterprise (BUM-Desa) Perancak Jembrana Bali [8].

2. Mangrove forest conservation

Conservation of natural resources and their ecosystems in principle consists of 3, namely (1) Protection of life support systems. In this case, it is important for the existence of flora and fauna and their ecosystems to receive protection, whether in a National Park area, nature reserve, wildlife reserve or community forest, customary forest; (2) preserving the diversity of plant and animal species and their ecosystems; (3) sustainable use of living natural resources and their ecosystems, which can play a role in the interests of science, research, education and training, culture, recreation and nature tourism and ecotourism. Flora and fauna conservation efforts aim to: (a) prevent plant and animal species from being endangered; (b) maintain genetic purity and species diversity; and (c) maintaining the balance and stability of the existing ecosystem; so that it can be used for human welfare in a sustainable manner (Republic of Indonesia Law No. 5 of 1990 [9]; Regulation of the Minister of Environment and Forestry No. P.106 of 2018 [10]).

Mangrove ecosystems have three main functions, namely physical/chemical functions, biological/ecological functions and economic functions. The economic function of mangrove forests is more directed at recreational tourism activities, educational tours and research tours. Furthermore, mangroves are managed for ecotourism activities. Mangroves play an important role in protecting coastal areas and maintaining habitat for a large number of species of animals/fauna (fishes, crustaceans, mollusks, reptiles, birds, to mammals), endangered species and endangered species, all of which play an important role in maintaining biodiversity.

The diversity of mangrove plant species plays an important role for the existence of mangrove fauna. The diversity of mangrove plants consists of two groups, namely true mangrove species and associated mangrove species. True mangrove species are plants that grow exclusively in mangrove habitat, including *Rhizophora* sp., *Bruguiera* sp. *Sonneratia* sp. *Xylocarpus* sp., *Avicennia* sp. and species of mangrove associations are terrestrial plants that are able to grow adaptively in mangrove areas, including *Hibiscus tiliaceus*, pescapre (*Ipomoea pes-capre* [1]).

Various types of fauna associated with mangrove ecosystems, namely various types of birds, insects and primates that live in the tree canopy as well as various types of fauna that live at the bottom of the mangroves such as wild boars, monitor lizards, crocodiles, snakes, shrimp, fish, shellfish, snails, crabs are an ecotourism attraction in mangrove ecosystems. Some of the bird species found in the mangroves of the Perancak Estuary include great egrets (*Egretta alba*), small egrets (*Egretta garzetta*), egrets (*Ardeola speciosa*) which are interesting attractions for ecotourism visitors [11].

Biodiversity conservation is important for several reasons, namely; (1) Ecological reasons. Individual species and ecosystems have developed over millions of years into complex dependencies. The greater the loss of habitat and species, the greater the danger of total collapse; (2) Economic reasons. Loss of biodiversity in general also means that species with economic and social potential may disappear before they are discovered; (3) Ethical reasons. When forests and other habitats are lost or degraded, so are the traditions and livelihoods of local people based on these habitats; and (4) esthetic reasons. Everyone would agree that a vegetated area with all its life content would be more interesting than a burnt, degraded landscape or large concrete buildings. Human existence is linked to the natural world. Every type of plant and animal is different from each other and this gives beauty to nature in different ways [12, 13].

The development of development in various sectors (including the tourism sector) has an impact on the environment, both the geophysical-chemical environment, the environment and the socio-culture of the surrounding community. This requires humans to always act wisely towards the environment so as not to cause negative or damaging impacts. In supporting programs to improve the management of living natural resources and their ecosystems in a harmonious, balanced and sustainable manner, various conservation efforts are required both in-situ and ex-situ to flora and fauna, especially to species that have been protected or are experiencing population decline. Flora conservation efforts are not only the responsibility of the government but also the wider community, non-governmental organizations, private institutions including entrepreneurs in the tourism sector.

Mangrove forests developed as mangrove tourism objects by governmental or non-governmental institutions based on ecotourism. Ecotourism activities are in principle beneficiary mangrove area while maintaining the biological/ecological functions of mangrove forests, there is a sustainable economic value and empowerment of local communities. The concept of ecotourism can be described in more detail in the principles of ecotourism, namely: (1) Minimizing physical, social, behavioral, psychological impacts; (2) Build environmental awareness, culture and respect; (3) Provide a positive experience for visitors and hosts; (4) Providing direct financial benefits for environmental conservation or preservation; (5) Generate financial benefits for local communities, private industry (6) Provide impressive interpretive experiences for visitors to increase sensitivity to the political, environmental, social climate of the tourist destination; (7) Build, operate facilities or infrastructure by minimizing environmental impact; (8) Recognizing the rights, spiritual beliefs of indigenous communities and empowering them [3, 4, 14].

The use of mangroves for ecotourism is in accordance with the development directions of the Sustainable Development Goals (SDGs), namely goal 12, regarding sustainable patterns of consumption and production; goal 13, on urgent action to combat climate change and its impacts; goal 14, regarding the conservation and sustainable use of sea, ocean and maritime resources for sustainable development; goal 15, on Protecting, restoring and promoting sustainable use of terrestrial ecosystems, managing forests in a sustainable manner, combating desertification, and halting and reversing soil degradation and halting the loss of biodiversity; and goal 17, on strengthening implementation measures and revitalizing the global partnership for sustainable development [15].

3. Use of mangrove ecosystems for ecotourism in Bali: case study

This section describes the distribution and mangrove forests area in Bali and some examples of mangrove areas that have been developed for ecotourism by non-governmental organizations or the local government. Based on data from the Ministry of Environment and Forestry, the area of mangrove land in Bali Province reaches 2143.97 hectares (3067.71 Ha), which is distributed in southern Bali (TAHURA Ngurah Rai) covering 1373.5 ha, Mangrove Nusa Lembongan covering 202 Ha, The Perancak Estuary, which is located in Jembrana Regency, Bali has mangrove forests with an area of 177.09 ha, Gilimanuk bay covering an area of 265.92 Ha and Buleleng Regency covering an area of 1291.40 Ha [16] (**Figure 1**).

The use of mangrove forests for ecotourism activities in several areas in the mangrove areas of the Province of Bali is based on management by local community groups/communities around the mangrove area and continues to maintain the conservation of biodiversity, landscapes and their ecosystems. Ecotourism activities carried out by this community group are fostered by the relevant agencies in their area and or state-owned enterprises which are part of their CSR (Corporate Social Responsibility) program.

Data on the diversity of flora, fauna and ecotourism attractions in the study area is the result of the author's observations and from several sources of articles that have been published in journals or books including [5, 17–22].

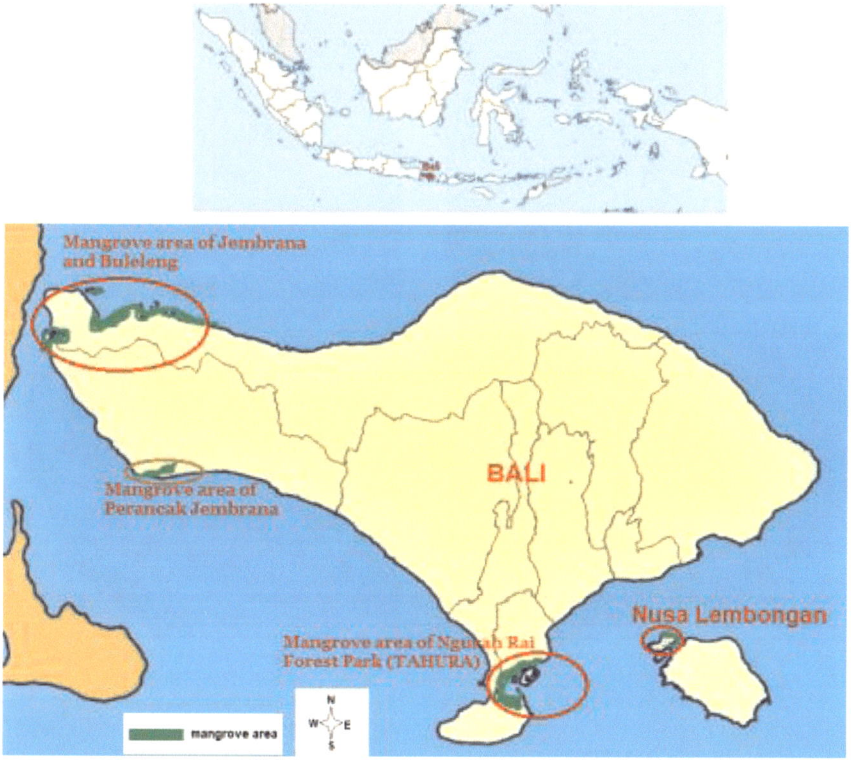

Figure 1.
Map of mangrove distribution in Bali (2021).

3.1 Mangrove ecotourism of Segara Guna Batu Lumbang

The Segara Guna Batu Lumbang mangrove forest is part of the Grand Forest Park (TAHURA) Ngurah Rai mangrove conservation area, which was developed by a community group, namely KUB Segara Guna Batu Lumbang. Geographically it is located at coordinates 8°44′4.02"S and 115°11'17.15"E. Ecotourism activities by this community group are fostered by the Forestry and Environment Service TAHURA Ngurah Rai and PT. Indonesia Power (Bali Power Generation Unit).

The mangrove forest in Segara Guna Batu Lumbang is a tourist attraction based on the conservation of the diversity of mangrove flora and fauna. Mangrove tours that are being developed are mangrove tours with canoes/jukung, fishing tours, volunteer tour, and spiritual tours. Mangrove tourist facilities in the Lumbang stone area, namely canoes, "jukung", boats, post/canoe base hall, fishing lines. The Segara Batu Lumbang mangrove tourism object combines spiritual tourism and conservation tourism (Voluntourism). In this area, the *Penyawangan Melasti* temple, Pemogan traditional village, Denpasar, Bali, canoe, boat route, stilt houses, posts for fishermen group facilities and facilities for tourists to be able to explore existing spots or objects (**Figure 2**).

Biodiversity in this mangrove area mainly consists of mangrove vegetation and mangrove association plants, bird fauna, mollusks, crustaceans, fish, reptiles, amphibians and some insects. Several species of plants and fauna are endangered species based on the IUCNRedlist, most of them are in the LC (Least Concern)

Figure 2.
Segara Guna Batu Lumbang mangrove tourist object (photos 2022).

No	Species	Common names	Conservation status
1	*Rhizophora mucronata*	Asiatic mangrove	LC, decreasing
2	*Rhizophora apiculata*	red mangrove	LC, decreasing
3	*Excoecaria agallocha*	blind-your-eye mangrove or molky mangrove	
4	*Xylocarpus granatum*	cannonball mangrove	LC, decreasing
5	*Sonneratia alba*	apple mangrove	LC, decreasing
6	*Bruguiera gymnorrhiza*	oriental mangrove	LC, decreasing
7	*Lumnitzera racemosa*	white-flowered black mangrove	LC, decreasing
8	*Aegiceras floridum*	black mangrove or river mangrove	NT, decreasing
9	*Avicennia marina*	gray mangrove or white mangrove	LC, decreasing

Scarcity NT: near threatened; LC: Least Concern, based on IUCN Red List, version 2022 [23].

Table 1.
Diversity of mangrove plants in Segara Batu Lumbang.

category. This shows that the mangrove area of Segara Guna Batu Lumbang has high conservation value (**Tables 1** and **2**). Expanse of mangrove forest and interacting fauna become an interesting sight for tourists visiting the Segara Guna Batu Lumbang mangrove tour (**Figures 3** and **4**).

Tourists visiting this area can travel around the mangroves by canoe, get to know the diversity of flora and fauna, research, bird watching, fishing, environmental education, become volunteers in mangrove conservation. The economic value

No	Species	Common names	Conservation status
	Birds		
1	*Gerygone sulphurea*	Golden-bellied gerygone	LC, decreasing
2	*Aegithina tiphia*	Common iora	LC
3	*Alcedo coerulescens*	Small blue kingfisher	LC
4	*Todiramphus chloris*	Collared kingfisher	LC, decreasing
5	*Alcedo menintin*g	blue-eared kingfisher	LC, decreasing
6	*Collocalia linchi*	Cave swiftlet	
7	*Bubulcus ibis*	Cattle egret	LC
8	*Egretta intermedia*	Intermediate egret	LC
9	*Egretta garzetta*	Little Egret	
10	*Egretta alba*	greater egret	LC
11	*Ardeola speciosa*	Javan Pond-heron	LC
12	*Ardea sumatrana*	great-billed heron	LC
13	*Butorides striata*	Striated heron	LC, Decreasing
14	*Orthotomus sepium*	olive-backed tailorbird	LC, Endemik Indonesia
15	*Streptopelia chinensis*	Spotted dove	
16	*Corvus macrorhynchos*	large-billed crow	LC
17	*Dicrurus macrocercus*	black drongo	LC
18	*Lonchura punctulata*	scaly-breasted munia	LC
19	*Hirundo tahitica*	Pacific Swallow	LC
20	*Merops philippinus*	Blue-tailed bee-eater	LC
21	*Nectarinia jugularis*	Olive-backed sunbird	LC
22	*Microcarbo melanoleucos*	Little Pied Cormorant	LC
23	*Phalacrocorax sulcirostris*	Little Black Cormorant	LC
24	*Pycnonotus goiavier*	yellow vented bulbul	
25	*Amaurornis phoenicurus*	white breasted waterhen	LC
26	*Rhipidura javanica*	Pied Fantail	LC, L
27	*Numenius phaeopus*	Eurasian whimbrel	LC, Decreasing, L.
28	*Tringa glareola*	sandpiper	LC
	Crustacean		
29	*Thalassina* sp.	Mangrove shrimp	
30	*Alpheus* sp.	Pistol shrimp	
31	*Sesarma roberti*	Sesarmid crabs	
32	*Uca annulipes*	fiddler crab	
33	*Cardisoma carnifex*	Crabs	
35	*Uca vocans*	fiddler crab	
36	*Scylla* sp.	Mangrove crab	
37	*Uca dussumieri*	fiddler crab	
38	*Uca triangularis*	fiddler crab	

No	Species	Common names	Conservation status
39	*Uca lactea*	fiddler crab	
40	*Coenobita* sp.	Hermit crabs	
	Mollusks		
41	*Assiminea brevicula*	Red mangrove snail	
42	*Cassidula nucleus*	Marine snail	
43	*Cerithidea cingulata*	Girdled horn snail	
44	*Chicoreus capucinus*	mangrove murex	
45	*Littorina scabra*	mangrove periwinkle	
46	*Nerita maxima*	nerites	
47	*Nerita planospira*	nerites	
48	*Neritina turrita*	nerites	LC
49	*Telescopium telescopium*	Telescope snail	
50	*Terebralia palustris*	the giant mangrove whelk	
51	*Terebralia sulcata*	sulcate swamp cerith	
52	*Geloina erosa*	Lokan mussel	
53	*Gelonia scabra*	Mangrove mussel	
54	*Placuna ephipium*	Saddle oyster	
	Fish		
55	*Glossogobius circumpectum*	Glosogobius fish	
56	*Periophthalmus barbarus*	mudskipper	LC
57	*Gazza minuta*	Toothed ponyfish	LC
58	*Scatophagus argus*	Spotted scat	LC
59	*Lates calcarifer*	Barramundi	
	Insects		
60	*Eurema* sp.	Yellow butterfly	
61	*Leptosia nina*	Small butterfly	
62	*Valanga nigricornis*	Javanese grasshopper	
63	*Orthetrum sabina*	green marsh hawk dragonfly	
64	*Pantala flavescens*	globe wanderer dragonfly	LC
65	*Crocothemis servilia*	scarlet skimmer dragonfly	LC
	Amphibian		
66	*Duttaphrynus melanostictus*	Frog	
	Reptile		
67	*Eutropis multifasciata*	Common Sun Skink	
68	*Cerberus rynchops*	dog face water snake	LC
69	*Varanus salvator*	Water monitor	LC

Scarcity, LC: Least Concern, base on IUCN Red List, version 2022 [23]; L: protected status according to Regulation of the Minister of Environment and Forestry No. P.106 of 2018 [10].

Table 2.
Diversity of fauna in the Segara Batu Lumbang mangrove area.

Figure 3.
Fishing and traveling around the mangroves at the Segara Guna Batu Lumbang, mangrove tourism attractions (photos: 2022).

Figure 4.
Birds in the Segara Guna Batu Lumbang mangrove area (photos: 2022).

generated from mangrove tourism activities is partly utilized for the welfare of the managing community group and partly for mangrove ecosystem recovery funds due to the impact of utilization (among them activities of planting mangrove seedlings, repairing facilities, funds for cleaning plastic waste).

3.2 Mangrove tourism of Segara Luhur Batu Lumbang

The Segara Luhur Batu Lumbang mangrove forest is part of the Tahura Ngurah Rai mangrove forest area. Geographically it is located at coordinates 8°43′33.86“S and 115° 12’1.27”E. This mangrove ecotourism was developed by the Simbar Segara group and the Manager of the Dalem Temple of Luhur Segara Batu Lumbang, Pemogan Village, Denpasar Bali. Ecotourism activities by this community group are fostered by the Forestry and Environment Service UPTD TAHURA Ngurah Rai. Guidance is carried out to ensure that the use of mangroves for tourism activities still prioritizes the conservation of flora and fauna and mangrove ecosystems.

The mangrove forest in Segara Batu Lumbang is one of the tourist attractions based on the conservation of the diversity of mangrove flora and fauna. The Segara Luhur Batu Lumbang mangrove tourism object combines spiritual tourism and conservation tourism (Voluntourism). An alternative type of tourism that has the opportunity to be developed in the Batu Lumbang Mangrove Forest is Voluntourism. This type of tourism combines volunteer activities and tourism. Because this area has beautiful natural potential [22] (**Figure 5**).

In this area, Pura Luhur Segara Batu Lumbang was built, by the traditional village of Pemogan, Denpasar, Bali. Mangrove tours that are being developed are mangrove tours with canoes, fishing tours, volunteer tours, and spiritual tours. Mangrove tourist facilities in this area, namely canoes, traditional boat "jukung", boats, post/canoe base hall, and fishing lines (**Figure 6**).

The diversity of flora and fauna in the Segara Batu Lumbang mangrove tourist area is almost similar to the flora and fauna in the Segara Luhur Batu Lumbang mangrove area, because it is still a TAHURA Ngurah Rai area. The flora and fauna consist of mangrove vegetation and mangrove association plants, bird fauna, mollusks, crustaceans, fish, reptiles, amphibians and some insects. Several species of plants and fauna are endangered species based on the IUCN Redlist, most of them are in the LC (Least Concern) category, there are several bird species which are protected species based on the Decree of the Minister of Environment and Forestry of the Republic of Indonesia no P106 of 2018. This shows that the Segara Batu Lumbang mangrove area has a high conservation value. The expanse of mangrove forest and interacting fauna is an interesting sight for tourists visiting the Segara Batu Lumbang mangrove tour. Tourists exploring mangroves using canoes are presented with a unique view of mangrove vegetation and several types of animals that interact with mangrove habitat (**Figure 7**).

Figure 5.
Segara Luhur Batu Lumbang tourism object (photos: 2022).

Figure 6.
Tour around the mangroves by canoe in the Segara Luhur Batu Lumbang mangroves (photos: 2022).

Figure 7.
Flora Fauna in the Segara Luhur Batu Lumbang mangrove (photos 2021 and 2022).

3.3 Mangrove ecotourism of Kampoeng Kepiting

The Kampoeng Kepiting mangrove ecotourism area is part of the Tahura Ngurah Rai mangrove forest area. Geographically it is located at coordinates 8°44′38.62“S and 115°11'0.03"E. This mangrove ecotourism was developed by the Nelayan Wanasari Group, Tuban Badung Village, Bali. Ecotourism activities by this community group are fostered by the Ngurah Rai Pertamina Depo (CSR Program) and the Forestry and Environment Service Tahura Ngurah Rai. Coaching is carried out for business development and ensuring that the use of mangroves for tourism activities still prioritizes the conservation of flora and fauna and mangrove ecosystems (**Figure 8**).

The uniqueness of the mangrove flora and fauna and the ecosystem in the Kampoeng Kepiting area is an attractive attraction for ecotourists. Ecotourism

Figure 8.
Mangrove ecotourism of Kampoeng Kepiting (photo: 2021).

Figure 9.
Voluntourism in Kampoeng Kepiting mangrove (photos: 2021).

attractions in Kampoeng Kepiting include mangrove tour packages using traditional boats, mangrove tours using canoes, fishing mangroves with traditional boats, crab aunt release tours, volunteer tours (planting mangrove seeds and cleaning mangroves from plastic waste), educational tours of mangrove ecosystems (**Figure 9**).

3.4 Mangrove tour tourism Nusa Lembongan

The mangrove forest in Nusa Lembongan which covers 202 ha [24] has been utilized by the community for mangrove tour tourism activities. Several tourism organizations that have developed a mangrove tour program in Nusa Lembongan include the Bali Tours Club, the Jungut Batu village mangrove tour group, Travelfish.org, the Tangjung Sanghyang tour group. In this activity, traveling tourists explore the mangrove forest by using rowing canoes, motorized canoes, some are via a trail. Throughout the tour, tourists are accompanied by local guides to enjoy the beauty of the expanse of mangrove forests.

Figure 10.
Mangrove tour in the mangrove forest of Nusa Lembongan (photos: 2022).

One of the community groups developing a mangrove tour in Nusa Lembongan is the Sari Segara Mangrove Tour Group. The mangrove area developed by this group is located at coordinates 8°40′1.25“S and 115°28’1.79”E. This group utilizes several parts of the mangrove forest area in Nusa Lembongan, specifically the mangrove forest in the northern part of Nusa Lembongan. The facilities used for the mangrove tour are canoes, traditional boats or canoes. Tourists can go around the mangroves while observing the diversity of flora and fauna (**Figure 10**).

Nusa Lembongan's mangrove forests support a diversity of mangrove plants, bird fauna, crustaceans, mollusks which are very interesting for tourists to enjoy. At least 11 species of true mangrove plants, 27 species of birds, 22 species of mollusks and 11 species of crustaceans have been recorded in the mangroves of Nusa Lembongan [5, 17, 25]. Most of the mangrove plant species are included in the rare LC, Vu and NT categories according to the IUCN Redlist. Likewise, most of the bird species found are endangered species, especially the LC category (**Tables 3** and **4**).

3.5 Ecotourism in mangrove ecosystem Pejarakan Buleleng

The coastal mangrove forest of Pejarakan Village covers 160 Ha, located at coordinates 8° 7′32.16“S and 114°34’19.89”E, managed by the Nature Conservation Forum Putri Menjangan (NCF Putri menjangan). Management of the area includes efforts to conserve mangroves and develop educational tours, ecotourism. The diversity of mangrove plants, growth zoning patterns, diversity of birds, mollusks, crustaceans that interact with mangrove ecosystems is an attraction for ecotourism [26].

The Nature Conservation Forum, which is a local community organization, is developing this area for ecotourism-based tourism. Facilities developed: office,

No	Species name	Common Names	Conservation status
1	*Rhizophora mucronata*	Asiatic mangrove	LC, decreasing
2	*Rhizophora apiculata*	Red mangrove	LC, decreasing
3	*Rhizophora stylosa*	spotted mangrove	LC, decreasing
4	*Excoecaria agallocha*	blind-your-eye mangrove or milky mangrove	
5	*Xylocarpus granatum*	cannonball mangrove	LC, decreasing
6	*Sonneratia alba*	apple mangrove	LC, decreasing
7	*Bruguiera gymnorhiza*	oriental mangrove	LC, decreasing
8	*Lumnitzera racemosa*	white-flowered black mangrove	LC, decreasing
9	*Ceriops decandra*	ten male mangrove	NT, decreasing
10	*Avicennia marina*	gray mangrove or white mangrove	LC, decreasing
11	*Avicennia lanata*	white mangrove	VU

Scarcity: LC: Least Concern, VU: Vulnerable, NT: near threatened, base on *IUCN Red List, version 2022, [23];* L: *protected status according to Regulation of the Minister of Environment and Forestry No. P.106 of 2018 [10].*

Table 3.
Mangrove plant species on Nusa Lembongan.

No	Species	Common names	Conservation status
	Birds		
1	*Alcedo coerulescens*	small blue kingfisher	LC, stable
2	*Amaurornis phoenicurus*	white breasted waterhen	LC, unknown
3	*Anthreptes malacensis*	brown-throated sunbird	LC, stable
4	*Ardea purpurea*	purple heron	LC, decreasing
5	*Butorides striata*	striated heron	LC, decreasing
6	*Cacomantis merulinus*	plaintive cuckoo	LC, stable
7	*Collocalia linchi*	cave swiftlet	LC, decreasing
8	*Copsychus saularis*	Oriental magpie-robin	LC, stable
9	*Gerygone sulphurea*	golden-bellied gerygone	LC, decreasing
10	*Hirundo tahitica*	Pacific Swallow	LC, unknown
11	*Hypothymis azurea*	black-naped monarch	LC, stable
12	*Lalage sueurii*	White shouldered triller	LC, increasing
13	*Lanius schach*	Long-tailed Shrike	LC, unknown
14	*Merops philippinus*	blue-tailed bee-eater	LC, stable
15	*Nectarinia jugularis*	live-backed sunbird	LC, stable
16	*Oriolus chinensis*	black naped oriole	LC, decreasing
17	*Pachycephala grisola*	mangrove whistler	—
18	*Passer domesticus*	house sparrow	LC, decreasing
19	*Psilopogon haemacephala*	coppersmith Barbet	LC, increasing
20	*Pycnonotus aurigaster*	Sooty-headed Bulbul	LC, decreasing
21	*Pycnonotus goiavier*	yellow vented bulbul	—
22	*Thalasseus bergii*	Greater Crested Tern	LC, stable

No	Species	Common names	Conservation status
23	*Todiramphus chloris*	collared kingfisher	LC, decreasing
24	*Todiramphus sanctus*	sacred kingfisher	LC, increasing
25	*Treron vernans*	pink-necked green pigeon	LC, stable
26	*Turnix suscitator*	barred buttonquail	LC, increasing
27	*Zosterops chloris*	lemon-bellied white-eye	LC, stable
	Mollusks		
28	*Anadara antiquata*	Shell fish	
29	*Anadara granosa*	blood cokle	
30	*Crassostrea rhizophorae*	The mangrove oyster	
31	*Modiolus micropterus*	horse mussel	
32	*Assiminaea sp.*	sea snail	
33	*Cerithidea obtusa*	mud creeper	
34	*Conus asiaticus*	cone snail	LC
35	*Conus virgo*	cone snail	LC
36	*Cypraea boivinii*	cowries	
37	*Cypraea caputserpentis*	snakehead cowry	
38	*Hemifusus ternatanus*	ternate false fusus	
39	*Indothais gradata*	rock snail/murex snails	
40	*Littorina carinifera*	Mangrove periwinkle	
41	*Littorina scabra*	mangrove periwinkle	
42	*Littoraria undulata*	mangrove periwinkle	LC
43	*Nerita plicata*	nerites	
44	*Nerita planospira*	nerites	
45	*Pugilina cochlidium*	spiral melongena	
46	Terebralia palustris	the giant mangrove whelk	DD
47	Terebralia sulcata	sulcate swamp cerith	
	Crustacean		
48	*Caenobita* sp.	hermit crab	
49	*Platypodia granulosa*	poisonous crabs	
50	*Scylla* sp.	mud crab	
51	Sesarma roberti	marsh /mangrove crab	
52	Sesarma sp.	marsh crab	
53	*Uca dussumieri*	Purple fiddler crab	
54	*Uca anulipes*	fiddler crab	
55	*Uca lactea*	fiddler crab	
56	*Uca tetragonon*	fiddler crab	
57	*Uca triangularis*	fiddler crab	
58	*Uca vocans*	fiddler crab	

Scarcity, LC: Least Concern, DD: data deficient, *(IUCN Red List, version 2022, [23])*.

Table 4.
Fauna species richness in the Nusa Lembongan mangrove.

information center, tracking/trail, resting point. Ecotourism activities can be in the form of education on mangrove ecosystems, mangrove conservation (nursery, planting), introduction to the diversity of flora and fauna, and research (**Figure 11**).

The diversity of mangrove plants and associated plants, the diversity of birds, mollusks, crustaceans and the typical landscape of mangrove ecosystems is an attraction for ecotourism attractions. Several true mangrove plant species have been recorded with particularities in root type, fruit shape, growth form and growing zoning in mangrove habitat. The species of mangrove species found include *Rhizophora apiculata, Sonneratia alba, Bruguiera gymnorhiza, aviccenia* sp. and *Rhozophora stylosa.*

In the Putri Menjangan mangrove area, 39 species of birds were found. The dominant species found were from the families Ardeidae and Alcedinidae, which are water bird species, including the gray herons (*Ardea cinerea*), sea herons (*Ardea sumatrana*), little silver egrets (*Egretta sacra*), silver egrets (*Egretta intermedia*), great egret (*E. alba*), kingfisher (*Halcyon chloris*), and blue shrimp (*Alcedo coerulescens*). The presence of these birds is an interesting attraction for bird watching ecotourism. The existence of mollusk and crustacean species is also an interesting attraction for ecotourism, several species of mollusks found include the canal monodont (*Monodonta canalifera*), giant mangrove whelk (*Terebralia palustris*), tropical periwinkle sea snail (*Planaxis sulcatus*), telescope snail (*Telescopium telescopium*), sea snail (*Cerithidea obtusa*), mud snail (*Cerithideopsilla alata*), periwinkle (*Littoraria strigata*). Several species from the crustacean group include sesarmid crabs (*Sesarma roberti*), Fiddler crabs (*Uca anulipes*), fiddler crabs (*Uca vocans*) and *Uca tetragonon* (**Figure 12**).

3.6 Perancak mangrove ecotourism

Coastal mangrove forest area of Perancak Jembrana village, which covers 10 hectares. Located at coordinates 8°23′55.16"S and 114°37'17.22"E. This mangrove

Figure 11.
Mangrove ecotourism facilities in Pejarakan mangrove (photos: 2017).

Figure 12.
Some interesting fauna that can be observed in pejarakan mangroves (photos: 2017).

ecotourism is managed by the Perancak Customary Village-Owned Enterprise (BUM-Desa) and under the auspices of the Jembrana Regency Government and the Ministry of Maritime Affairs and Fisheries, for Ecotourism. Ecotourism activities are based on three concepts, namely the preservation of the flora and fauna of the mangrove forest is maintained, there is a sustainable economic value and the local community plays a role in its management. In this area several facilities supporting ecotourism activities were built, including a wooden trail for tracking, an office for ticket reservations, ecotourism information boards and toilets. All facilities are built with an environmentally friendly concept [27] (**Figure 13**).

The expanse of the coastal mangrove vegetation of Perancak and the existing biota is an interesting attraction. Mangrove plant species with a distinctive root type (stilt

Figure 13.
Perancak mangrove ecotourism object. (photo source: Bali tripon.com. Accessed: 2022).

root, pneumatophor, knee root), various fruit shapes (ball-like, chili-like, heart-shaped, bean-shaped), unique growing habitat, namely in the intertidal area (tidal).

4. Conclusions

The conclusion is that by utilizing mangrove forests for ecotourism activities, there are several important things, namely:

1. Flora fauna, landscapes, ecological systems in mangrove forests are still running in balance.

2. The function of the mangrove forest is biologically, physically and chemically maintained.

3. There is a sustainable economic value for social and financial welfare for ecosystem recovery.

4. The mangrove forest is a research laboratory for researchers and students.

5. Mangrove ecotourism has value for environmental education for the visiting community, which in turn is disseminated to the community, government and entrepreneurs.

6. There is a change in perspective in the utilization of living natural resources, namely from exploitation to sustainable use.

7. Ecotourism Providing interpretive experience, good knowledge for visitors about the existence of flora and fauna, especially in the ecotourism area.

8. Ecotourism pays respect to regional and community religious values, local cultural wisdom, and the social life of local communities.

Acknowledgements

I would like to thank the heads of the mangrove tourism groups, namely Segara Guna Batu Lumbang, Segara Luhur Batu Lumbang, Kampoeng Kepiting, the heads of the NCF Putri Menjangan and the Sari Sagara Lembongan Group for their permission and facilitation during the field survey. We hope that the information in this book chapter will benefit the development of ecotourism, especially mangrove ecotourism in Bali.

Conflict of interest

I declare that what I wrote in this book chapter is purely for the benefit of developing positive information for sustainable mangrove management and there is no conflict of interest with other parties.

Author details

I. Ketut Ginantra
Biology Study Program, Faculty of Mathematics and Natural Sciences, Udayana University, Bali, Indonesia

*Address all correspondence to: ketut_ginantra@unud.ac.id

References

[1] Kitamura S, Anwar C, Chaniago A, Baba S. The development of sustainable mangrove management project. In: Handbook of Mangroves in Indonesia-Bali & Lombok. Indonesia: Ministry of Forestry Indonesia and Japan International Cooperation Agency; 1997

[2] Nybakken JW. In: Ediman TM, Koesoebiono DGB, Hutomo M, Sukardjo S, editors. Marine Biology an Ecological Approach. Jakarta: PT. Gramedia; 1988

[3] Sudarto G. Ecotourism. In: Nature Conservation Efforts, Sustainable Economic Development and Community Empowerment. Indonesia: The Kalpataru Bahari Foundation and the Indonesian Biodiversity Foundation; 1999

[4] Duangjai W, Tuntates U, Kroeksakul P. The comparative evaluation of community-based ecotourism Management at Mangrove Forest Communities in Satun Province, Thailand. International Journal of Emerging Technology and Advanced Engineering. 2017;**4**(6):42-48

[5] Ginantra IK, Suaskara IBM, Joni M. Diversity of mangrove Plant for Support Ecotourism Activities in nature conservation forum Putri Menjangan, Pejarakan Buleleng-Bali. Journal of Environmental Management and Tourism. 2018;**(Volume IX, Fall), 5** (29):987-994. DOI: 10.14505/jemt.v9.5 (29).10

[6] Wahyuni IGAS, Ardhana IPG, Sudirga SK, Ginantra IK. Plants As Ecotour attractions around trail of mangrove information Centre In mangrove Forest of southern Bali. Jurnal Bumi Lestari. 2006;**6**(2):1-8

[7] Sari IP, Yoza D, Sribudian E. Feasibility analysis of mangrove ecosystem as an ecotourism object in Teluk Pambang Village, Bantan District, Bengkalis District. Jurnal Faperta. 2015;**2**(1):1-10

[8] Bali Tripon.com. Perancak Bali Mangrove Ecotourism. Available from: https://www.balitripon.com/ekowisata-mangrove-perancak/. [Accessed: 2022]

[9] Republic of Indonesia Law No. 5. Concerning the Conservation of Living Natural Resources and their Ecosystems. Jakarta: Republic of Indonesia; 1990

[10] Regulation of the Minister of Environment and Forestry No. P.106. Concerning about the Species of Plants and Animals Protected. Indonesia Republic: Minister of Environment and Forestry; 2018

[11] Muhaerin M. Study of Mangrove Ecosystem Resources for Ecotourism Management in Perancak Estuary, Jembrana [Thesis]. Bali: Department of Water Resources Management, Faculty of Fisheries and Marine Sciences, Bogor Agricultural University; 2008

[12] Irwaantoshut.com. Biodiversity Conservation. Available from: http://www.irwantoshut.com/

[13] Forest Stewardship Council (FSC). High Conservation Value Forests (HCVFs). FSC; 1999. Available from: https://connect.fsc.org/. Accessed in 2022.

[14] Indonesia Ecotourism Institute. Principles of Ecotourism. 2022. Available from: https://www.indonesiaecotourism.com/

[15] Supriatna J. Biodiversity Conservation: Theory and Practice in Indonesia. Indonesia: Pustaka Obor Indonesia Foundation; 2018

[16] Prameswari AASG, Hariyanto T, Sidik F. Analysis of mangrove vegetation index using an alos avnir-2 satellite. Geoid. 2015;**11**(1):40-45

[17] Ginantra IK, Muksin IK, Joni M. Crab diversity as support for ecotourism activities in Pejarakan mangrove Forest, Buleleng, Bali, Indonesia. Biodiversitas. 2021;**22**(10):4139-4145. DOI: 10.13057/biodiv/d221003

[18] Ginantra IK, Muksin IK, Suaskara IDM, Joni M. Diversity and distribution of mollusks at three zones of mangrove in Pejarakan, Bali, Indonesia. Biodiversitas. 2020a;**21**(10):4636-4464. DOI: 10.13057/biodiv/d211023

[19] Ginantra IK, Muksin IK, Joni M. Diversity of birds for ecotourism attractions in the mangrove ecosystem of nature conservation forum Putri Menjangan, Buleleng Bali. Journal of Environmental Management and Tourism. 2020b;**XI, Spring, 1**(41):105-113. DOI: 10.14505//jemt.v11.1(41).07

[20] MacKinnon J, Phillips d K, Ballen VB. Birds in Sumatra, Java, Bali and Kalimantan (Including Sabah, Sarawak and Brunei Darussalam) [LIPI-Seri Panduan Lapangan]. Bogor: Puslitbang Biologi-LIPI; 2010

[21] Suta PWP, Mahagangga IGAO. Community-based tourism development (case study in Kampoeng Kepiting ecotourism, Tuban, Bali). Jurnal Destinasi Pariwisata. 2017;**5**(1):144-149

[22] Tyas S, Dan Arida INMS. Community-based Voluntourism development opportunities in Batu Lumbang mangrove Forest, Suwung, Pemogan, Denpasar. Jurnal Destinasi Pariwisata. 2020;**8**(2):2020

[23] IUCN Redlist. Version 2022. Threatened species. Available from: https://www.iucnredlist.org [Accessed: February 26, 2020]

[24] Palguna IBA, Ardhana IPG, Wayan Arthana IW. Mangrove forest structure and diversity in Nusa Lembongan, Nusa Penida Sub District, Klungkung Distric. Ecotrophic. 2017;**11**(2):108-115

[25] Ginantra IK, Muksin IK, Joni M, Yuni LPEK. Bird diversity as a support of ecotourism activities in the mangrove ecosystem of Lembongan Island Bali-Indonesia. Journal of Environmental Management and Tourism, (Volume XIII, Winter). 2022;7(63):1840-1850. DOI: 10.14505/jemt.v13.7(63).04

[26] NCF Putri Menjangan. Profile of NCF Putri Menjangan Bali, Indonesia. Pejarakan Village: NCF Putri Menjangan; 2016

[27] Jembrana District Government, Perancak Village Has Mangrove Forest Ecotourism. Available from: https://jembranakab.go.id/index.php?module=detailberita&id=3356